T0310644

Evolutionary Intelligence for Healthcare Applications

This book highlights various evolutionary algorithm techniques for various medical conditions and introduces medical applications of evolutionary computation for real-time diagnosis.

Evolutionary Intelligence for Healthcare Applications presents how evolutionary intelligence can be used in smart healthcare systems involving big data analytics, mobile health, personalized medicine, and clinical trial data management. It focuses on emerging concepts and approaches and highlights various evolutionary algorithm techniques used for early disease diagnosis, prediction, and prognosis for medical conditions. The book also presents ethical issues and challenges that can occur within the healthcare system.

Researchers, healthcare professionals, data scientists, systems engineers, students, programmers, clinicians, and policymakers will find this book of interest.

AIoT: Artificial Intelligence of Things – A Powerful Synergy
Series Editor: S. Balamurugan

The powerful synergy of Artificial intelligence (AI) and the Internet of Things (IoT) is now becoming a new horizon in the Industry 4.0 revolution. Despite the fact that the full ability of AI and the IoT are still in their baby steps, the powerful synergy of these two technologies could improve the problem-solving ability and could lead to transforming the world to the next level in automation. The capacity to house huge amounts of information has been rising in the last three to five years. This also comes along with privacy and security concerns. Sensors today can be applied to almost everything. This gives space for efficiently collecting huge amounts of real-time data and process the same instantaneously. The emergence of Artificial Intelligence based techniques like Machine Leaning and Deep Learning exhibits high capacity to solve complex problems. AI and IoT are independently exhibiting their powers in industrial applications, a super power could be tapped by synergizing AI and IoT (AIoT). The capability of IoT is expanded exponentially while applying artificial intelligence based strategies for efficient decision making. Forthcoming years will witness more optimization and networking opportunities by the synergy of Artificial Intelligence and IoT. It has been written in research reports by Industrial Verticals that complete human-computer integration will be achieved by the year 2030. Also by the synergy of AI and IoT and also on advances in robotics and automation, up to 40% percent of the current workforce could be replaced by technology within the next 10–15 years. Considering the aforementioned research predictions the need for this Focus book series becomes important to keep updated with the synergies in Artificial Intelligence and IoT.

Evolutionary Intelligence for Healthcare Applications
T. Ananth Kumar, R. Rajmohan, M. Pavithra, and S. Balamurugan

Evolutionary Intelligence for Healthcare Applications

T. Ananth Kumar, R. Rajmohan,
M. Pavithra, and S. Balamurugan

CRC Press
Taylor & Francis Group
Boca Raton London New York

CRC Press is an imprint of the
Taylor & Francis Group, an **informa** business

First edition published 2023
by CRC Press
6000 Broken Sound Parkway NW, Suite 300, Boca Raton, FL 33487-2742

and by CRC Press
4 Park Square, Milton Park, Abingdon, Oxon, OX14 4RN

CRC Press is an imprint of Taylor & Francis Group, LLC

Library of Congress Cataloging-in-Publication Data

Names: Kumar, T. Ananth (Tamilarasan Ananth), author. I Rajmohan,
R., author. I Pavithra, M., author. I Balamurugan, S. (Shanmugam), 1985-author.
Title: Evolutionary intelligence for healthcare applications / T. Ananth
Kumar, R. Rajmohan, M. Pavithra, S. Balamurugan.
Description: First edition. I Boca Raton : CRC Press, 2023. I Includes
bibliographical references and index.
Identifiers: LCCN 2022038908 (print) I LCCN 2022038909 (ebook) I
ISBN 9781032185040 (hardback) I ISBN 9781032185071 (paperback) I
ISBN 9781003254874 (ebook)
Subjects: MESH: Medical Informatics Applications I Artificial Intelligence
Classification: LCC R855.3 (print) I LCC R855.3 (ebook) I NLM W 26.55.A7 I
DDC 610.285--dc23/eng/20220919
LC record available at https://lccn.loc.gov/2022038908
LC ebook record available at https://lccn.loc.gov/2022038909

ISBN: 978-1-032-18504-0 (hbk)
ISBN: 978-1-032-18507-1 (pbk)
ISBN: 978-1-003-25487-4 (ebk)

DOI: 10.1201/9781003254874

Typeset in Times
by KnowledgeWorks Global Ltd.

Contents

Preface

In the recent two decades, the acceptance of metaheuristics, particularly swarm intelligence and evolutionary algorithms (EAs), has exploded. Researchers offer a plethora of algorithms and approaches every year, and they discover a growing number of innovative applications. When we consider our daily lives, it becomes apparent that intelligent systems play a greater part in our work or practical actions, and numerous changes have been noted over the past several years in terms of intelligent approaches, methods, and techniques. Currently, the discipline of metaheuristics is having considerable influence on the spectrum of healthcare, particularly in illness diagnosis and medicine discovery. Genetic intelligence has evolved as a new-generation technique belonging to the evolutionary computing category. While evolutionary computation, especially when biologically inspired, may be effective for search and optimization, selecting an optimal solution from a search space that is often huge and/or complicated is very much in line with the natural evolution process. Typically, the underlying process of evolution is driven by a stochastic heuristic that is suitable to a given optimization environment. In genetic intelligence, the search for an optimal in a search space is defined by the way a swarm approaches its objective.

The healthcare industry needs to use genetic intelligence and EAs to develop new drugs and improve the quality of care provided to patients around the world. This will help the industry thrive and grow in the long run by reducing mortality rates and improving quality of life for patients. Healthcare quality improvement and cost reduction are two of the most pressing challenges in today's healthcare business. Scheduling, planning, forecasting, and optimization challenges may all be described as evolutionary computation problems in the healthcare business. Even though evolutionary computation has been used to schedule and plan trauma systems and generic drug manufacturing, other problems in the healthcare industry, such as strategic planning in CAD prognosis and disease control, have not been properly developed for evolutionary information processing techniques.

The authors of this book were inspired to compile all EAs techniques and algorithms for healthcare, especially disease diagnosis, drug discovery, and bioinformatics modeling, to provide readers to research on this potential avenue of optimization in more contexts. The book would benefit several researchers and students as they can treat this as a strong base for exploring more opportunities of evolutionary intelligence algorithms in healthcare, and researchers can explore a lot of unexplored areas to find scope of problems to perform novel research. The book has been categorized into seven chapters to discuss applications of evolutionary computation in disease diagnosis, especially for heart, diabetes, degenerative, tumors, tuberculosis, muscular dystrophy, COVID-19, and bioinformatics.

Chapter 1 deals with investigations on Mathematical Models of Evolutionary Intelligence. It starts by discussing about findings of various researchers from a variety of scientific and technical areas who have been investigating the capability of EAs. It contains primitive representation and primitive inference. This chapter briefly described the conceptual methodology behind evolutionary intelligence algorithms. Moreover, the basic structural components of EAs and the various types of EAs are discussed. Finally, the role of EA in healthcare is analyzed and summarized with respect to futuristic needs.

Chapter 2 mainly discusses the diagnosis of heart diseases using evolutionary intelligence and modeling. It includes evolutionary architecture, challenges, motivation, and research directions, followed by optimization mechanisms in heart classification, mathematical schemes like genetic offloading techniques, and optimization schemes. Finally, applications of genetic algorithms are employed to predict the occurrence of heart disease. Morcover, the importance of genetic algorithm in the discovery of relevant features in heart diagnosis is analyzed.

Chapter 3 studies, examines, and analyzes the most current uses of EA technology to various elements of diabetes education. This study tries to give insight and recommendations for the development of prospective, data-driven decision support platforms for diabetes care, with an emphasis on tailored patient management and lifetime educational interventions, using the information and evidence collected.

Chapter 4 introduces the clinical treatments of degenerative diseases like Parkinson's and Alzheimer's. Next, the early prediction of neuro-degenerative disease and challenges in the diagnosis are discussed. Finally, the EAs for treating degenerative disorders are briefed.

Chapter 5 deals with tuberculosis disease diagnosis with evolutionary modeling. It discusses the challenges and issues in tuberculosis classification based on pulmonary and extrapulmonary problems. This chapter also provides a selection of examples chosen to illustrate how evolutionary intelligence is advancing the field of infectious illness and assisting institutions in combating them more effectively.

Chapter 6 mainly discusses about the examination and treatment of muscular dystrophy and emphasizes the healthcare team's involvement in treating individuals with this illness. This chapter describes the muscular characteristics, EA model selection (e.g., traditional machine learning models versus EA learning models), and feature engineering which are discussed as crucial components of creating dystrophy diagnosis systems. In addition, we stress the difficulties of such healthcare-oriented dystrophy systems and suggest many research prospects for the medical and computer science communities.

Chapter 7 discusses about diagnosis of brain tumor using nature-inspired computing. Evolutionary population-based optimization algorithms are prevalent optimization algorithms inspired by nature. The aim of this chapter discusses about both carcinoma classification and tumor diagnosis using evolutionary and genetics modeling.

Acknowledgments

I would like to thank the Almighty for giving us enough mental strength and belief in completing this work successfully. I thank my friends and family members for their help and support. I express my sincere thanks to the management of IFET College of Engineering, Tamilnadu, India. I wish to express my deep sense of gratitude and thanks to CRC Press for their valuable suggestions and encouragement.

T. Ananth Kumar, Ph.D.

I thank my family and friends for their prolonged support towards the successful completion of the book. I express my sincere thanks to the management of VIT-AP University, Amaravati, Andhra Pradesh, India. Also, I would like to thank the CRC Press for giving me the opportunity to edit this book.

R. Rajmohan, Ph.D.

I express my sincere thanks to the management of IFET College of Engineering, Tamilnadu, India, I would like to take this opportunity to specially thank CRC Press for kind help, encouragement and moral support.

M. Pavithra, M.Tech.

I thank my friends and family members for their support towards the successful completion of the book. I wish to express my deep sense of gratitude and thanks to CRC Press for their valuable suggestions and encouragement.

S. Balamurugan, Ph.D.

About the Authors

Dr. T. Ananth Kumar is working as Associate Professor in IFET college of Engineering affiliated to Anna University, Chennai. He received his Ph.D. degree in VLSI Design from Manonmaniam Sundaranar University, Tirunelveli. He received his master's degree in VLSI Design from Anna University, Chennai, and bachelor's degree in Electronics and communication engineering from Anna University, Chennai. He has presented papers in various National and International Conferences and Journals. His fields of interest are Networks on Chips, Computer Architecture, and ASIC design. He is the recipient of the Best Paper Award at INCODS 2017. He is the life member of ISTE and few membership bodies. He has eight patents in various domains. He has written many book chapters in Springer, IET Press, and Taylor & Francis press.

Dr. R. Rajmohan is currently working as Assistant Professor (Senior Grade-I) in VIT-AP University, Amaravati, Andhra Pradesh. He completed his Ph.D. in the field of wireless network at SSN College of Engineering under Anna University. He received his master's degree in Network and Internet Engineering from Pondicherry University, Pondicherry, and bachelor's degree in Computer Science and Engineering from Pondicherry University, Pondicherry. He has published more than 70 papers in various reputed SCI and Scopus-indexed journals. His fields of interest are Wireless Network, Deep learning, and IoT. He has won the best educator award from the International Institute of Organized Research (I2OR) in the year 2019. He is the lifetime member of various educational bodies and acted as reviewer for Springer and other standard journals. He is currently acting as the academic editor for PLOS ONE journal.

Ms. M. Pavithra received her master's degree in Distributed Computing System from Pondicherry University, Puducherry. She received her bachelor's degree in Computer Science and Engineering from Anna University, Chennai. She is working as Assistant Professor in IFET College of Engineering afflicted to Anna University, Chennai. Her areas of interest are Artificial Intelligence, Network Security, and Computational intelligence.

Dr. S. Balamurugan PhD, SMIEEE, ACM Distinguished Speaker is the Director of Albert Einstein Engineering and Research Labs. India. He received his B.Tech. degree from PSG College of Technology, Coimbatore, India, M.Tech., and Ph.D. degrees from Anna University, India. He has published 60 books, 300 international journals/ conferences, and 200 patents. He is also the Vice-Chairman of the Renewable Energy Society of India (RESI). He is also serving as a research consultant to many Companies, Startups, SMEs, and MSMEs. He is the series editor of book series' and serving in various editorial capacities of several International Journals. He is also the recipient of the Young Scientist Award, Certificate of Exceptionalism, Outstanding Scientist Award, and Best Director Award. His biography is listed in "Marquis WHO'S WHO," United States. His research interests include Artificial Intelligence, Augmented Reality, Internet of Things, Big Data Analytics, Cloud Computing, and Wearable Computing. He is a life member of ACM, IEEE, ISTE, and CSI.

Evolutionary Intelligence

1

1.1 INTRODUCTION

Evolutionary intelligence (EI) is popular in their field because it helps researchers and scientists improve their work. These techniques also perform admirably when applied to diagnosing medical conditions and creating new medications. There are many complex problems in the healthcare field, the majority of which cannot be solved in polynomial time [1]. The forecasting and diagnosis of fatal diseases could be greatly aided by techniques derived from the field of evolutionary computation (EC). Therefore, medical students exert significant effort to implement these strategies in clinical and biomedical settings. This study's primary objective is to provide an overview of the various EI methods and algorithms currently employed in the healthcare and medical fields for disease diagnosis. EI is beneficial not only to clinicians, counselors, and medical doctors but also to those who work in technical roles within the health industry, such as medical physicists and technicians, and those who want to learn more about systems in order to either implement them or better understand them [2]. As a result, a concise introduction to EI is required for disease diagnosis in numerous healthcare and medical fields and specialties. The subsequent chapters in this book will each have an introduction similar to this one.

DOI: 10.1201/9781003254874-1

1.2 PRELIMINARIES

1.2.1 Evolutionary Computation

Using artificial intelligence, a group of global optimization algorithms known collectively as "evolutionary computation" were developed. In a wide range of computational models, evolutionary computation (EC) techniques are implemented as essential design components. These models utilize evolutionary processes as their primary method of problem-solving; consequently, the techniques of EC are a fundamental aspect of their design. The natural evolutionary process that occurs in the world was used as a model for evolution and optimization in this scenario. EC, which is inspired by the process of biological evolution, can be used to find solutions to difficult optimization problems. In the field of healthcare application development, these techniques are commonly employed as analyzers and prediction algorithms. Similar techniques are employed in resolving complex problems, such as those with a large number of constraints or variables [3]. According to Darwin's theory, a population's ability to survive can be increased through the mutation of its existing members. This is made possible by natural selection. According to numerous pieces of evidence, the only individuals with a greater probability of having children are those who live longer. In other words, longer-lived individuals have a greater chance of having their genetic characteristics and specific traits selected from the population. Computers model the events that occur during the evolutionary processes of a variety of natural phenomena to convert these rules and apply them with the aid of computation to enhance their capacity for knowledge acquisition and the development of tools. EC is a collection of powerful techniques based on Darwinian evolution that is used to achieve a variety of functions, such as optimization, learning, and self-adaptation [4]. EC is a collection of numerous potent techniques that are based on Darwinian evolution and are used to accomplish a variety of tasks.

1.3 EVOLUTIONARY ALGORITHMS

Several components comprise the design of an evolutionary algorithm (EA), including representation, parent selection, crossover operators, mutation operators, survival selection, and termination conditions [5]. These are some of the components that an EA may contain.

1.3.1 Representation of EA

First, it is essential to represent the genotypes and then map them to the phenotypes before proceeding to the representation step. In general, designers try to maintain as much consistency and condensability in the representations of genotype and phenotype as possible. This is done so that measurement metrics, like distance, can be mapped to phenotype space metrics without the loss of any information that is semantic [6].

A few examples of representations of EAs are shown in Figure 1.1: (a) representation of an integer, (b) representation of the structure of a protein on a lattice model, and (c) representation of a mathematical expression using a tree.

When it comes to the processing of information, an EA can, in general, make use of a wide variety of data representations. Tree structures, linear structures with variable lengths, and fixed- and variable-length linear structures are all examples of different kinds of linear structures. This figure displays three different illustrations for your perusal. A vector consisting of integers is depicted in Figure 1.1a. It has been noticed that the array of binary genes that make up its genotype consists of 10 different elements. Figure 1.1b [7] illustrates the relative encoding representation of a protein based on the Hydrophobic-Polar (HP) lattice model. [Note: Figure 1.1a depicts the absolute encoding representation of a protein.] Figure 1.1c is a diagram that illustrates a mathematical expression in the form of a tree. This tree structure is obviously a structure that can have a variable length and a flexible design.

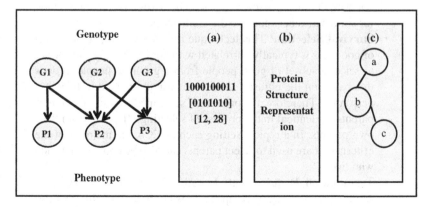

FIGURE 1.1 Examples of evolutionary algorithms.

1.3.2 Components of Evolutionary Algorithms (EAs)

The various components of the EA [8] can be classified into the following categories:

- **Selecting Parents:** This step aims to locate healthy parents that can be utilized in subsequent crossings. The level of goodness a parent demonstrates is directly proportional to how physically fit that parent is. The vast majority of parent selection programs are successful because they provide more opportunities to the parents who are more physically fit than to the other parents. Because of this, there is a greater chance that "good" children will be born.
- **Operators of Change:** It is comparable to the process that occurs when an individual or animal gives birth. Because of this, these two categories of operators, in addition to mutation operators, are collectively referred to as "reproductive operators." The majority of the time, a crossover operator will combine the genes of two different individuals to create the genetic makeup of a third individual. It endeavors to disassemble a person into their parts before reassembling those parts into an entirely new individual.
- **Mutation Operators:** This component simulates the mutation mechanism, which is the natural process by which certain genetic components evolve over time. As a result, a mutation operator will alter various components of an individual's genome in the process of performing a standard modeling practice. On the other hand, mutations can be seen as a way to look for new things, which is the opposite of what crossover operators do. Crossover operators look for similarities between two different genes.
- **Survival Selection:** This technique assumes that a person's level of goodness is typically correlated with how physically fit they are to select a subset of good people from a group of people. It does this based on the phrase "goodness is usually related to how fit a person is." Therefore, the process of selecting offspring from their parents is identical to the process of selecting offspring from their own parents. In a typical setting such as EC4, the majority of the criteria that are used to select parents can also be applied to decide who lives.
- **Termination Requirement:** In order for an evolutionary algorithm to be considered complete, it is necessary for this requirement to be satisfied.

1.3.3 What Varieties of EA Are There to Choose from?

The formula for the genetic programming algorithm. The genetic algorithm has been around longer than any other EA that has been studied. The idea was founded on Darwin's theory of evolution, which served as the foundation. In today's world, distinguishing between a genetic algorithm and an EA is becoming an increasingly difficult task [9]. Individuals frequently use the phrases "evolutionary algorithm" and "genetic algorithm" when discussing different types of computer programs. In order to explain the operation of a genetic algorithm in a way that is simple to comprehend, we decided to use the canonical genetic algorithm [10] as a representative example. In the standard implementation of the genetic algorithm, the phenotype of each individual is represented by a binary array of fixed length. This array contains the person's genotype. After that, the normalized chance of being chosen is calculated by dividing each individual's fitness by the group's average fitness as a whole. The algorithm then utilizes the data to determine which parents should participate in a one-point crossover to produce offspring that are subsequently subjected to mutations. In the end, those individuals who are born as a result of reproduction are the ones who go on to live in the subsequent generation, and so on.

Genetic programming is a subset of genetic algorithms that can be studied on its own or in conjunction with other genetic algorithms. How they present themselves differentiates them from one another. When it comes to genetic programming [11], genotypes are the trees that can either stand for expressions or programs (Figure 1.1 depicts an example). Both parent selection and survival selection in genetic programming can, to a limited extent, still make use of some of the standard strategies for selection that are utilized by EAs. However, this is only to a certain degree. Because genetic programming languages use crossover and mutation operators, these languages are distinguished from other programming languages. For instance, exchanging subtrees between two distinct trees and creating subtrees by accident are both excellent ideas.

Differential evolution is a heuristic method for optimizing globally continuous space functions that are not linear and cannot be differentiated. This method was developed in the 1960s. The term "evolutionary differentiation" refers to differential evolution on occasion. There are many different kinds of evolutionary computing algorithms, including the differential EA. There are still a great many more. In the same way that other well-known direct search methods, such as genetic algorithms and evolution strategies, begin with a set of potential solutions, the differential EA [12] does the same thing. These candidate solutions are improved through iteratively introducing mutations into the population and selecting only the best of those that produce a lower value for the objective function to keep in the population. This procedure will continue until the most appropriate response has been determined.

Evolution Strategies, also known as ESs, is a type of direct search (and optimization) method that is modeled after how nature functions. They are a subset of a larger group of algorithms that are collectively referred to as "evolutionary algorithms" (EAs). These methods gradually improve solutions by subjecting a population of individuals with candidate solutions to mutation, recombination, and selection [13]. Former EAs took many more cues from natural events than the current evolutionary strategy. Instead, it was fabricated in a laboratory to be utilized as a tool for numerical optimization. Because of this, the structure of the evolution strategy is entirely dissimilar to that of other EAs.

Multimodal optimization is a type of optimization that entails finding all or the majority of the locally best solutions to a problem. This can be done in order to maximize performance. This contrasts the conventional optimization method, which centers on identifying a single optimal solution. One of the subfields that fall under the umbrella of EC is known as "evolutionary multimodal optimization" [14]. In various ways, this subfield is connected to machine learning. It is also possible to look at multiple solutions to the optimization problem at hand in order to discover previously unknown relationships or properties. It is essential to have multiple solutions to acquire domain knowledge because these solutions can be analyzed to discover previously unknown things. In addition, the vast majority of algorithms for multimodal optimization not only locates multiple optimal solutions in a single iteration but also maintain the diversity of their populations, enabling them to perform global optimization on multimodal functions.

1.3.4 Typical EA Pseudo Code

The way that nature operates inspired the development of EAs. It is typical for EAs to start with a population that has been seeded at random at the very beginning of the process. After that, it takes several generations for another shift in the population. When a new generation is starting, the people who will become the parents are selected because they are healthy. They produce new individuals through the process of reproduction, which results in offspring. These completely new human beings are brought into the world. Generally speaking, people will refer to this process as the generation process. Before beginning this procedure, a sample of children at random was chosen to experience a series of transformations. The algorithm will then use the survival selection criteria to determine which individuals have the most excellent chance of successfully passing their genes to the subsequent generation.

Algorithm 1: Evolutionary Algorithm

Choose suitable representation methods;

> A(s): Population time for Parent s
> B(c): Population time for Offspring c
> S□ 0;
> Initialize A(s);
> while not termination condition do
> temp = Parent Selection from A(s); B(c + 1) = Crossover in temp;
> B(c + 1) = Mutate B(c + 1);
> if overlapping then
> A(s + 1) = Survival Selection from B(c + 1) U A(s); else
> A(s + 1) = Survival Selection from B(c + 1);
> end if
> s □ s + 1;
> end while
> Good individuals can then be found in A(s)

This will be done by determining which individuals have the highest probability to accomplish this; we will determine which individuals have the highest probability of escaping unharmed. According to [15], if the algorithm described overlaps, both the parent and offspring populations will be considered when determining who will live and who will die. Only the parent population will be considered if the algorithm does not overlap. This is the reason for it, and it has to do with the complexity of overlapping algorithms. The algorithm needs to take into account both sets of inputs to consider both groups properly. In this scenario, the children are the only ones who will participate in the process of natural selection for survival. Only those individuals who have been accorded preferential treatment are permitted to pass on their lineage to subsequent generations. This privilege is not extended to the general population. Only those people who have been accorded special consideration are eligible to pass on their lineage to the subsequent generation. It is necessary to carry out this process several times that is not specified to achieve the goal [16].

1.4 ROLE OF EA IN HEALTHCARE

For the most part, medical decisions can be described as a search for the proper location. An expert pathologist who examines biopsies for malignant cells will search through all possible cell features to find a collection of features that will enable him or her to make a conclusive diagnosis [17]. Radiologists search for the treatment that will be the most effective among all the possible treatments when planning a series of radiation doses [18].

Healthcare-related web searches tend to be comprehensive and in-depth. Using clinical tests as a foundation for making decisions in the field of medicine is a smart move. Only one conclusion can be drawn after considering all of these points of evidence (e.g., benign or malignant). Searching can be difficult because variables in the domain are interdependent, and many real-world problems are non-linear. This makes it difficult to find similar things across the search space. This is due to the inherent non-linearity of the vast majority of problems encountered in the real world and the close interdependence between the domain's variables. In order to test and evaluate various machine learning techniques, a wide variety of medical issues are used as benchmarks [19]. Utilizing EC, it is possible to design effective search strategies for large, complex areas.

Evolving techniques make use of natural mechanisms to find the best way to search every area possible. The likelihood of getting stuck in an optimal local solution is reduced because of the parallelism they possess. Using the data currently available, it is difficult to build accurate models of the decision-making process in medicine. There are too many non-linear and uncertain parameters in the models to treat them analytically on the one hand. Medical specialists are frequently unavailable or unwilling to collaborate when it comes to turning their expertise into a useful decision-making tool [20].

This is even though many medical databases are accessible to the general public. Nowadays, medical results are both an archive of the patient's medical history and an important repository for medical knowledge, thanks to the widespread adoption of electronic databases. Thus, the outcomes are able to fulfill both purposes at once. Increasingly sophisticated computational processing tools are required to make effective use of the data that is currently available. Since there is so much data, it is difficult to keep up with it all. EC is used to perform a wide range of functions in medicine. When making a medical decision, it is almost always possible to find a place for evolutionary techniques to be applied. Medical imaging and signal processing, as well as planning and scheduling, are examples of jobs that EAs perform in medicine [21, 22].

1.5 CONCLUSION

Computer science and the natural world are brought closer together through EAs. The EA places a greater emphasis on gaining knowledge from observing natural processes than it does on developing original ideas. New computer techniques are incredibly versatile and can be applied to a wide variety of contexts, and their foundation or model can be found in the natural laws that govern the universe. These algorithms have a wide range of potential applications in the medical field, including the diagnosis of conditions such as cancer and brain tumors, diabetic retinopathy, and cardiovascular disease, as well as the design and development of medications by the pharmaceutical industry. This chapter examines the fundamental structure and components of EAs, the various types of EAs, how they are utilized, and the reasons why they are significant to the healthcare field.

REFERENCES

1. Chauhan, R. S., Taneja, K., Khanduja, R., Kamra, V., & Rattan, R. (Eds.). (2022). *Evolutionary computation with intelligent systems: A multidisciplinary approach to society 5.0.* Boca Raton, FL: CRC Press.
2. Lu, Y. (2019). Artificial intelligence: a survey on evolution, models, applications and future trends. *Journal of Management Analytics*, 6(1), 1–29.
3. Glover, F. (1986). Future paths for integer programming and links to artificial intelligence. *Computers and Operations Research, 13*, 533–549.
4. Nayyar, A., Garg, S., Gupta, D., & Khanna, A. (2018). Evolutionary computation: Theory and algorithms. In *Advances in swarm intelligence for optimizing problems in computer science* (pp. 1–26). New York, NY: Chapman and Hall/CRC.
5. Reddy, S. (2018). Use of artificial intelligence in healthcare delivery. In Heston, T. (ed.). *eHealth-making health care smarter* (pp. 1–16). IntechOpen. https://doi.org/10.5772/intechopen.74714
6. Zhou, Z. H., Yu, Y., & Qian, C. (2019). *Evolutionary learning: Advances in theories and algorithms* (pp. 3–10). Singapore: Springer.
7. Krasnogor, N., Hart, W., Smith, J., & Pelta, D. (1999). *Protein structure prediction with evolutionary algorithms. International genetic and evolutionary computation conference (GECCO99)* (pp. 1569–1601). Portland, OR: Morgan Kaufmann.
8. Ma, X., Li, X., Zhang, Q., Tang, K., Liang, Z., Xie, W. ... & Zhu, Z. (2018). A survey on cooperative co-evolutionary algorithms. *IEEE Transactions on Evolutionary Computation, 23*(3), 421–441.
9. De Jong, K. A. (2006). *Evolutionary computation: A unified approach.* Cambridge, MA: MIT Press.

10. Wong, K. C., Leung, K. S., & Wong, M. H. (2010). Effect of spatial locality on an evolutionary algorithm for multimodal optimization. In *EvoApplications 2010, Part I, LNCS 6024* (pp. 481–490). Berlin, Germany: Springer-Verlag.

11. Banzhaf, W., Trujillo, L., Winkler, S., & Worzel, B. (2022). *Genetic programming theory and practice XVIII*. East Lansing, MI: Springer Nature.

12. Opara, K. R., & Arabas, J. (2019). Differential evolution: a survey of theoretical analyses. *Swarm and Evolutionary Computation, 44*, 546–558.

13. Wierstra, D., Schaul, T., Glasmachers, T., Sun, Y., Peters, J., & Schmidhuber, J. (2014). Natural evolution strategies. *The Journal of Machine Learning Research, 15*(1), 949–980.

14. Wong, K. C. (2015). Evolutionary multimodal optimization: A short survey. *arXiv preprint arXiv:1508.00457*.

15. Whitley, D. (1994). A genetic algorithm tutorial. *Statistics and Computing, 4*(2), 65–85.

16. Chui, K. T., Lytras, M. D., Visvizi, A., & Sarirete, A. (2021). An overview of artificial intelligence and big data analytics for smart healthcare: requirements, applications, and challenges. *Artificial Intelligence and Big Data Analytics for Smart Healthcare*, 243–254.

17. Walters, H. (2019). The evolving role of the radiologist. *Future Healthcare Journal, 6*(2), 93.

18. Malik, H., Iqbal, A., Joshi, P., Agrawal, S., & Bakhsh, F. I. (Eds.). (2021). *Metaheuristic and evolutionary computation: algorithms and applications*. Singapore: Springer Nature.

19. Shailaja, K., Seetharamulu, B., & Jabbar, M. A. (2018, March). Machine learning in healthcare: A review. In *2018 Second international conference on electronics, communication and aerospace technology (ICECA)* (pp. 910–914). Coimbatore, India: IEEE.

20. Cai, C. J., Reif, E., Hegde, N., Hipp, J., Kim, B., Smilkov, D. … & Terry, M. (2019, May). Human-centered tools for coping with imperfect algorithms during medical decision-making. In *Proceedings of the 2019 chi conference on human factors in computing systems* (paper 4, pp. 1–14). New York, NY: Association for Computing Machinery. https://doi.org/10.1145/3290605.3300234

21. Freitas, A. A. (2003). A survey of evolutionary algorithms for data mining and knowledge discovery. In *Advances in evolutionary computing* (pp. 819–845). Berlin, Germany: Springer.

22. Nolle, L., & Schaefer, G. (2009). Evolutionary computing and its use in medical imaging. In *Computational Intelligence in Medical Imaging: Techniques and Applications* (pp. 41–60). Chapman and Hall/CRC.

Heart Disease Diagnosis

2

2.1 INTRODUCTION

The human heart beats with a force that is greater than that produced by any other human muscle. The heart's position within the chest is very close to the anatomical center of the chest. When it is closed, the human adult heart has about the same circumference as a fist. The human heart beats between 75 and 80 times per minute on average. This equates to approximately 115,000 occurrences every single day and 42 million occurrences every year. Throughout a typical adult's lifetime of 70 years, the human heart will have completed more than 2.5 billion beats [1]. Even when the body is at rest, the heart continues to pump blood to various body parts. Heart rates typically range between 60 and 100 beats per minute in adults and children older than 10 years. An athlete who has put in the necessary work to improve their fitness should have a heart rate between 40 and 60 beats per minute. The right atrium, which collects blood from veins and pumps it to the right ventricle, is one of the four chambers that make up the heart. The other three chambers are the left ventricle, left atrium, and left atrium. After receiving blood from the right atrium, the right ventricle pumps this blood to the lungs, where it becomes oxygen-rich. After this process, the right ventricle is responsible for regulating blood pressure. It is the job of the left atrium, which is situated on the left side of the heart, to pump oxygenated blood from the lungs to the left ventricle, which is located on the opposite side of the heart.

The left ventricle of the heart is the most potent chamber and is responsible for pumping oxygen-rich blood throughout the body. The forceful contractions of the left ventricle are responsible for the production of our blood pressure. Coronary arteries are blood vessels that deliver blood rich in oxygen to the heart's muscle. These arteries can be found running along the surface of the heart. In addition, the heart is made up of a network of nerve tissue

DOI: 10.1201/9781003254874-2

responsible for the transmission of complicated signals that control the contraction and relaxation of the heart muscle. The pericardium is a sac that wraps the entire heart from all sides. Heart valves are like doors between your heart chambers. They do this by moving back and forth to control the volume of blood allowed to pass through. Blood can move freely between the upper and lower chambers of the heart thanks to a set of valves known as the atrioventricular (AV) valves. The tricuspid valve is the name given to the valve that sits between the right atrium and the right ventricle of the heart.

The mitral valve is located in the door between left atrium and left ventricle. When blood is pumped out of the ventricles, the semilunar (SL) valves relax and allow blood to flow freely. The aortic valve is the valve in the heart that opens when the left ventricle relaxes, allowing blood to flow freely from the left ventricle into the aorta (the artery that carries oxygen-rich blood to your body). When the pulmonary valve is allowed to relax, it allows unrestricted blood flow from the right ventricle to the pulmonary arteries (the only arteries that carry oxygen-poor blood to your lungs). The term "cardiovascular disease," which is synonymous with "heart disease," (HD) refers to a collection of conditions that adversely affect the cardiovascular system, specifically the heart and blood vessels. These diseases can affect a single chamber of the heart, multiple chambers of the heart, or the blood vessels themselves. A person can show disease symptoms depending on their health state. The term "heart disease" refers to a collection of conditions affecting the heart and the blood vessels.

2.2 HEART ATTACK

Myocardial infarction, more commonly known as a heart attack, is at the top of the list of cardiovascular diseases affecting people in the United States, according to both statistics and personal stories. In the United States, on average, someone has a heart attack about every 40 seconds. Even though heart attacks are often shown in movies and on TV, the symptoms you can see on the outside don't tell you what is going on inside your body. A heart attack happens when the heart's muscle is suddenly cut off from the oxygen it needs to work correctly. This happens when the blood flow that is used to bring oxygen to that part of the body slows down or stops. This is because of a condition called atherosclerosis, which is caused by the slow buildup of plaque in the coronary arteries. This is something that can cause a heart attack. Plaque is made up of fatty deposits, cholesterol, and various other things. Blood clots can form around the plaque. This can slow down or stop the flow of blood, which can cause a heart attack.

Stroke is usually thought of as a type of HD because it changes how blood flows. This is because of how two things are related to each other. On the other hand, a stroke isn't caused by problems with the heart. Instead, it's caused by problems with how blood gets to the brain. Ischemic strokes make up 87 percent of all strokes and are caused when a blood vessel that brings blood and oxygen to the brain gets blocked. If the condition isn't treated as soon as it's noticed, some parts of the brain could be hurt or even die if they do not get enough blood and oxygen. Hemorrhagic strokes can also be caused by other things, like when blood vessels in the brain grow out of shape or when there are problems with the blood vessels.

2.2.1 Heart Attack

Heart failure is a condition that happens when the heart cannot pump blood as well as it should. Congestive heart failure is another name for heart failure. Even though the condition's name might lead to believe that the heartbeat has stopped completely, that is not the case. Even though the heart is still beating, it is not pumping blood at a fast enough rate for the body to keep functioning normally. Heart failure that isn't treated can cause symptoms like tiredness and shortness of breath, making even simple things like walking or climbing stairs hard.

2.2.2 Arrhythmia

Arrhythmia of the heart is a general term for any abnormal heartbeat, such as one that is too slow, too fast, has an irregular beat or tempo, or a mix of these things. If the heart does not have the right rhythm, it cannot pump as much blood to the rest of the body. There is a chance that the heart will not be able to pump enough blood to get enough oxygen and nutrients to the other organs in the body.

2.2.3 Heart Value Complications

Heart value complications can include a wide range of heart valve problems. Stenosis is a condition where the heart valves don't open all the way to let blood flow typically. This makes it hard for the heart to work right. Regurgitation is a condition that can happen when the valves of the heart do not close properly, letting blood leak backward into the heart's chambers. Since this is the case, blood will flow backward through the heart. The heart's valves and arteries must be in good shape to lower the risk of complications that could cost a life.

2.2.4 Hypertension – Heart Disease

When a person's high blood pressure is the leading cause of cardiovascular disease, the person's blood pressure rises to levels that are medically concerning. High blood pressure makes the heart work less well, making it less able to do its job. These problems can be dangerous to a person's life. High blood pressure can cause several life-threatening conditions, such as heart failure, thickening heart muscle, damage to the coronary arteries, and many others.

This chapter consists of evolutionary intelligence (EI) to explain the various types of HD. Doctors' difficulties in diagnosing HD are also discussed, as the impact of EI on this research.

2.3 HEART DISEASE CLASSIFICATION USING EA

HD is the most complex and dangerous problem to deal with on a global scale right now. Because of this, the heart cannot pump enough blood to the rest of the body. This can lead to heart failure, a common problem for people with HD. Many things make it hard to diagnose and treat HD, especially in countries with insufficient diagnostic equipment or trained medical staff. Because of this, cardiology patients can get better diagnoses and treatments. The most important parts of the HD analysis process are choosing the disease classification and the features to be looked at [2]. HD is put into groups with the help of a high-dimensional dataset. High dimensionality can mess up the HD classification procedure, but you can avoid this by using the normalization method. So, the normalization method reduces the number of dimensions, and the mean-mode replacement method fills in the gaps with a new value. With the Filter, Wrapper, and Embedded strategy, it was possible to choose and remove important features. An evolutionary algorithm (EA) has been used to classify things successfully. Figure 2.1 shows the whole organizational structure of the HD classification system.

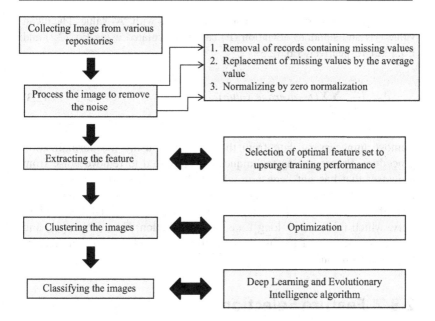

FIGURE 2.1 Organizational structure of heart disease classification.

2.3.1 Preprocessing

The first step in the HD diagnostic process is the preparation of the data [3]. The fact that the dataset has unnecessary data and information that was left out contributes to the decreased accuracy of the classification. The average residual value is calculated by using Formula 1, and the value that was left out of the calculation is found and filled in (1).

$$PS = \frac{\sum_{j=0}^{t} Xz(j)}{m} \tag{2.1}$$

The data from the heart is represented by the letter "m" in Eq. (2.1). When referring to "average value," is mentioned as "PS." To normalize the HD dataset, the zero-normalization technique must be used, which involves calculating the mean and standard deviation of the entire dataset. Using the Equation, the process can be completed (2):

$$A' = \frac{A - H(y)}{F(Y)} \tag{2.2}$$

It is possible to represent the normalized data with "A" value. The mean value (H) and standard deviation (F) can be calculated using Eq. (2.1) and Eq. (2.2), respectively Eq. (2.3),

$$\partial = \sqrt{\frac{1}{m} \sum_{j=1}^{m} X\left(t - average\ value\right)^2} \tag{2.3}$$

Remove unnecessary data from the hard drive using the normalization procedure. Preprocessing techniques are employed to remove noise from a dataset that has not been normalized. This is done to improve classification performance, which noise has been shown to impact negatively. Classification is slowed down due to the features of high dimensionality, which results in a long time for a prediction. Filter, Wrapper, and Embedded are just a few methods available for reducing the number of feature dimensions.

2.3.2 Feature Selection

The process of selecting features, also known as feature extraction, is an essential component of pattern recognition and machine learning (ML). Because of the feature selection (FS) methods, the computation cost is reduced, and there is also the potential for improving classification performance [4].

2.3.3 Filter Methods

In order to determine whether or not a filtering method is successful, cross-validation is not utilized. Instead, univariate statistics are used to assess the overall quality of the features that are being measured. These methods are superior to those involving wrappers in terms of both speed and the amount of effort required. Filtering methods are more effective when working with datasets with a high dimension count.

Let's discuss a few of these different approaches:

- **Information Gain:** The degree to which the entropy of a dataset is reduced as a result of changes made to that dataset. It is referred to as the information gain. You can use it to choose which features to use by determining the information gained from each variable in relation to the variable you want to use it for. You can use this method to choose which features to use.

- **Fisher's Score:** When someone is watching, one of the most common ways to choose features is based on what is called the Fisher's score. The algorithm we will use will provide us with the ranks of the variables in descending order, and these ranks will be based on the Fisher's score. After that, we can select the variables based on what is currently taking place.
- **Correlation Coefficient:** The term "correlation" comes from the field of statistics and refers to a method for determining the degree to which two or more variables are connected linearly. We can determine what will happen with the first variable by comparing it to two others and seeing how they behave. The selection of features based on correlation is predicated on the assumption that essential variables will closely connect to the objective being sought. Additionally, a connection must exist between the variables and the target, but there must be no connection between the variables themselves.

When two variables are connected in this way, we are able to make predictions about the behavior of one variable based on the behavior of the other variable. Since this is the case, the model only needs to consider one of the related features if there are two of them, given that the other feature does not contribute any new information.

2.3.4 Wrapper Methods

Wrappers require a method capable of searching the space of all possible subsets of features and evaluating the quality by learning and evaluating a classifier using only those features. A specific ML algorithm directs the selection of features, which is then used to optimize a given dataset by applying it to the selected features. It accomplishes this by employing a technique known as greedy searching, which involves evaluating every possible feature combination in light of the evaluation criterion. In the vast majority of instances, the predictive accuracy of the wrapper methods is significantly higher than that of the filter methods.

2.3.5 Forward Feature Selection

Determine which variable has the highest level of performance relative to the objective as the initial step in this iterative methodology. After that, we select a second variable that will produce the best overall performance results when added to the initial variable. This procedure is repeated numerous times until the desired quality level is achieved.

2.3.6 Backward Feature Elimination

This technique operates in a manner that is opposed to the forward FS technique. In this phase, we will construct a model by compiling a list of accessible features. Then, utilizing this data, we extract from the model the variable that provides the most precise value for the evaluation metric. This process is repeated multiple times until the desired level of performance is achieved.

2.3.7 Embedded Methods

These methods incorporate interactions between characteristics while maintaining an acceptable computational cost [5]. This allows them to reap the benefits of both the wrapper and filter methods. Embedded methods are considered iterative because they are responsible for managing each iteration of the process of training the model and painstakingly extracting the features that contribute the most to training for each iteration. This indicates that embedded methods are iterative in the conventional sense.

2.3.8 LASSO Regularization (L1)

In ML, preventing a model from becoming overfit involves applying a penalty to each of the model's parameters. This technique is also known as regularization. In order to restrict the model's freedom, this is performed. In the regularization of linear models, the penalty is applied to the coefficients that multiply each predictor. So that the model remains linear, this is performed. This demonstrates that the model is as precise as possible. One of the many types of regularization that can reduce coefficients to zero is the lasso regularization, also known as the L1 regularization. Consequently, there is no longer a need to include this component in the model.

2.3.9 Random Forest Importance

The random forest algorithm is a bagging technique that combines the outputs of numerous decision trees into a single output. Random forests rank the tree-based strategies they employ based on how effectively they improve the node's purity, or, in other words, a reduction in the impurity (Gini impurity) across all trees. The last nodes of trees contain notes that have undergone the slightest improvement in purity, while the very first nodes contain notes that have undergone the most significant improvement in purity. As a result, we are able to create a subset of the essential characteristics by pruning the branches of the trees below a specific node in the hierarchy.

2.4 CHALLENGES AND ISSUES IN HEART DISEASE DIAGNOSIS

More hospitals and other healthcare facilities are opening as people become more aware of how vital healthcare is and as technology improves. On the other hand, it is still hard to find healthcare that is both affordable and of high quality. Chronic diseases are a big problem for public health because they cause half of all deaths worldwide. Non-infectious diseases are also the leading cause of death and the leading cause of healthcare costs in the United States. People over 65 years are more likely to have HD, which has recently surpassed infectious diseases as the leading cause of death worldwide. Since these diseases are becoming more common, more people are getting sick. This makes everyone's healthcare costs and social costs go up. Because of this, it is essential to take the proper precautions.

2.4.1 Traditional Systems

Doctors have a harder time making quick decisions when there are a lot of ambiguous factors to consider. Classification systems help doctors quickly and accurately analyze medical data. The first step is to create a model that can be used to classify existing records based on sample data. Classifiers for HD diagnosis have been developed using a variety of algorithmic approaches [5].

There is no other way to stop HD than to find and treat its underlying causes. This means that a solution can be found in the appropriate techniques are employed. Therefore, it is critical to discover what causes HD and develop a rapid, accurate method of diagnosis [6]. Medical diagnostic systems need to be feature-selective to accurately diagnose patients and predict their long-term health [7]. Cutting-edge diagnostic techniques are required to identify this disease accurately.

Pattern recognition and ML have the same problem: deciding which features to include in their datasets. Extracting features from a dataset is known as "feature extraction" (ML). In some cases, FS methods improve classification performance while reducing the amount of time spent on the computer. One of the most pressing issues in ML and data mining is how to represent data effectively. Some features can be used to classify or predict, but not all of them. Several of the dataset's features are either unnecessary or redundant and, therefore should be removed. The classification may not be as accurate as it could be because of this feature. The FS process is recommended for classification and regression problems in order to improve the classifier's overall performance and reduce the amount of work it needs to do [8].

New measurement techniques have made it possible to collect medical data that includes both useful and useless data and information that has already been collected. The description of the target class deteriorates when features that aren't critical to the target class are included. It doesn't help to include redundant characteristics when describing a target class, as they only make the process more cumbersome. To find the target class, some features are used multiple times. It is necessary to thoroughly search the sample space in order to extract useful information from these datasets [9].

Because the heart is the body's engine and regulates blood flow, HD is hazardous and can even be fatal [10]. According to the World Health Organization (WHO), anxiety is a leading cause of death worldwide. HD is expected to claim the lives of 17.9 million people worldwide by the year 2021, according to current projections. A person's life can be saved if HD is detected early. There are currently a lot of studies underway to find reliable methods for doctors to diagnose a wide range of illnesses. A smaller number of factors (also referred to as "attributes") that are more closely associated with cardiac disease are examined in this study, which makes use of classification to predict diagnosis more accurately.

2.4.2 Existing Methodologies for Diagnosing Heart Diseases

Accuracy was achieved by using linear kernel support vector machine (SVM) classifiers, which Ashish et al. used in their detection of HD [11]. The method presented in [12] for hybrid neural networks had a reported accuracy of 96.8 percent. This method was demonstrated. The algorithms' separability split value, k-nearest neighbor (KNN), and feature space mapping were used in the study [13] to detect HD. Compared to the other two algorithms, KNN performed the best in classification accuracy (85.6 percent).

HD can be diagnosed using the method outlined in reference [14]. Particle swarm optimization (PSO) and neural network feed-forward back-propagation are used in this method. A decision tree is one of the tools used in the HD data mining procedure described in [15]. Using data mining techniques to detect cardiovascular disease was investigated by researchers [16]. Different classification methods, such as neural networks and decision trees, are used to forecast HD and identify the factors that are most likely to play a significant role in its progression. To overcome the difficulties they encountered, the researchers in this study employed a wide range of research methods.

This repository at the University of California, Irvine (UCI) contains a large number of studies that have evaluated the classification prediction accuracy of the numerous clustering and classification algorithms that have been applied to HD [15]. You can find this information on the UCI website. In the information environment, several attempts have been made to make better decisions and recommendations using evidence. This is because better analytical methods for predicting chronic HD are urgently required. The provision of appropriate medical advice, backed up by an accurate forecast of the patient's likelihood of developing a minor illness in the near future, is one of the most important roles played by healthcare systems. It's critical to understand that medical research [13] includes a variety of models for predicting the likelihood of developing various diseases.

Researchers have spent much time trying to find the most accurate ML method to investigate the links between HD and other factors. This article aims to predict and correct HD diagnoses, which aims to avoid unintended errors, reduce medical costs, and improve overall treatment quality [12].

Therefore, the researchers have presented statistical methods for understanding the three medical datasets to produce prediction models by extracting relevant diagnostic rules. They did this in order to deal with the previously mentioned problems [15]. Decision trees, Naive Bayes (NB), SVM, and a priori algorithms have been used in this investigation to produce satisfactory results for the researcher. In the study described in citation [17], a fuzzy system supported by a genetic algorithm was used to try and predict the likelihood of developing HD. When it comes to predicting cardiovascular disease, the fuzzy decision support system (FDSS) method proposed has a high level of performance.

Artificial neural network (ANN), classification and regression tree (CART) algorithm, neural network, and logistic regression have been found to have respective accuracy levels of 97 percent, 87.6 percent, 95.6 percent, and 72.2 percent [18]. And the accuracy of the statistical technique known as logistic regression has been determined to be 97 percent. Reference [19] states that for two-, three-, and four-class classification problems, the accuracy of an automated method for early detection of class changes in patients with heart failure using classification algorithms was 97.87 percent and 67 percent, respectively. On a dataset that included evaluation and validation approaches, the method was tested. In order to achieve this level of precision, researchers used an evaluation validation strategy dataset in conjunction with the algorithms they developed. In order to investigate a wide range of classification issues using a wide variety of classification algorithms, this dataset has been used by many researchers [20].

Decision trees were proposed as an alternative classification method by Ghiasi et al. [21]. When compared to the current methods, this method yields superior results to the others. To incorporate data mining into various disease forecasting initiatives currently underway, the proposed method was devised in the medical field. Using a decision tree-based method and neural system classifiers for cardiovascular disease diagnosis and prognosis were both successful (HD). For coronary artery disease prediction, data mining outperformed all other approaches. Healthcare administrators found using a decision tree to aid in the diagnosis of coronary disease to be a boon.

The study's authors, Nagavelli et al. [22], created a diagnostic system based on ML to predict HD risk in patients. There were seven different ML methods and three different FS strategies used to conduct the performance evaluations. As a result, it is easy to distinguish between those who have and those who don't have HD, as well as classify and identify them. Additionally, the optimistic receiver curves and the area under the curve for each classifier were analyzed. Using this decision support system, doctors are able to diagnose heart patients accurately.

Balamurugan et al. [2] demonstrated that Balamurugan and his colleagues could process the presence of HD features. The FS process was developed by combining an artificial gravitational cuckoo search (AGCS) algorithm with a particle bee optimized associative memory neural network (PBAMNN). In order to better mimic the behavior of AGCS algorithms, the feature was reduced in size. This was done to ensure a better match. Associative memory classifiers can now be used to process the data after the features have been discovered. Analyzing the results of simulations helped determine the usefulness of the proposed method.

The author [23] proposed a method for predicting chronic HD in patients using ML methods and Fourier transformations. As a result of this procedure, an effective medical recommendation system was created. In this context, "input" can refer to either the data or the time series information about the patients. The Fourier transformation can be used to extract information about frequencies from this. NB, ANN, and SVM were all used as classifiers in this study, as were ANN. The study's findings show that using the deep learning methodology to produce reliable and accurate recommendations for cardiac patients was a success.

Improved HD diagnosis has been made possible by Brites and colleagues [24]. For the sake of FS, this procedure must be followed. HD's recognition as a diagnostic category in the medical field was made possible by data science. A variety of HD classification strategies had already been developed, but the prediction accuracy was still lacking. A decision was made that simulations using multiple HD datasets would be the best option for the author to look into. For the sake of accuracy, this was implemented.

An ideal HD diagnosis method was presented by Magesh and Swarnalatha [25], that used cluster-based design thinking (DT) learning to select features (CDTL). It is possible to divide the CDTL into five distinct stages. Based on the distribution of the target labels, the initial dataset was divided into its component parts. The samples with a wide distribution across the population were used to create the other possible group combination. As a result of the use of entropy, we discovered the commonalities between each class-set combination. Additional acute characteristics have been included in the development of an entropy divider. Using all the noteworthy features, the HD diagnosis performance was successfully completed.

Tawid et al. were able to improve their ability to diagnose HD in patients by combining the rough sets (RSs) and PSO with the transductive support vector machine (TSVM) [26]. The data was located in the UCI repository at the time of this retrieval. The Z-score method was selected as the basis for the procedure used to normalize the data because it was believed to increase the data's reliability while also reducing the amount of redundant data. Choosing the most advantageous attributes was the primary goal of the preliminary phase of implementing the PSO based on the RS-based attribute reduction method. The system's overall performance is boosted, while the complexity of the computations required is raised. After that, the RBF-based TSVM classification method was used to classify HD (radial basis function).

Kaur et al. [17] created an adaptively optimized fuzzy logic classification model for HD classification. This section will cover both the fuzzy classification model and the RS-based FS module. Here, an alternative method for selecting the most valuable characteristics is presented: the "roughly based" technique. An essential part of the genetic algorithm is the generation of new chromosomes, fitness evaluation, crossover, mutation, and selection. Fuzzy rules are part of the classification optimization strategy, and some examples include fuzzification, fuzzy rule generation, and defuzzification. Fuzzy rule generation also uses the term "fuzzification." Various high-dimensional datasets from UCI's ML repository were used to verify the experiments. These datasets included the Swiss, Hungary, and Cleveland datasets.

Koppu et al. [27] used robotics and deep learning to develop a model for predicting HD. Early disease detection is a significant goal of this research, as it allows humans the opportunity to live longer and healthier lives. The cleaning procedure was the primary concern, in this case, to ensure that any missing values could be removed. The principal component analysis has been used to conceptualize the process of eliciting characteristics. Choosing the traits with the most significant fitness potential was a dragonfly algorithm (FODFA) job. A deep belief network (DBN), also known as a DBN, was used to obtain the feature vector. To test the model's performance, three datasets were used: the Wisconsin breast cancer dataset (WBC dataset), the Cleveland dataset (Cleveland HD dataset), and the Starlog HD dataset (HD dataset). The development and implementation of an optimized neural network model for diagnosing and classifying HD in an extensive data framework [28]. The online and offline prediction stages of the TLB-ANN model's operation can be distinguished from one another because of how it operates. The UCI repository provided the data needed to conduct an experimental evaluation.

Individual classification algorithms used to detect HD can benefit from FS methods because of these effects. A patient's risk of developing HD can be estimated using one of these algorithms. Noise and dependency relationships in the dataset can hinder the diagnosis process. There are many records of symptoms that have been recorded more than once in the initial datasets. There are also many records about accompanying syndromes. This necessitates the use of a technique known as FS to eliminate features that are redundant or merely extraneous from the initial feature set. Various classifier strategies for HD classification have been developed and documented in many academic papers. These methods contribute to an improvement in the HD diagnosis system's general level of functionality. When trying to determine whether or not HD classification is effective, the research community is up against a formidable obstacle because of the observations mentioned above. The research above identifies a few limitations for the classification of HD from the dataset and makes some predictions regarding the most recent method for classifying HD from the dataset.

2.5 EA FOR HEART DISEASE DIAGNOSIS

Evolutionary modeling for HD diagnosis is shown in Figure 2.2. FS can be used to reduce the dataset's dimensionality, which in turn can be used to create a model. In order to do this, only some of the measured features are selected (predictor variables). One of the most important aspects of statistical learning is reducing the data's dimensionality or the number of features. A large number of features but a small number of observations in datasets, such as those used in bioinformatics, is typically not useful for producing

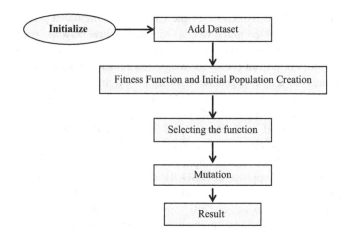

FIGURE 2.2   Evolutionary modeling for heart disease diagnosis.

the desired learning result and the small number of observations may cause the algorithm to overfit to the noise. Many of these features are ineffective in achieving the desired learning outcome. The data used in bioinformatics is an example of a collection. Reduce the number of features in a system to save storage space, speed up computations, and make it easier to understand.

For this reason, GA is particularly well-suited to problems in which domain knowledge and theory are either unavailable or impractical to address. Using the GA, the best solution to a problem is found by searching for it until a specific criterion is met. This will be done by conducting a thorough search. In order to achieve this result, the selection, crossover, and mutation operators should all be used at the same time.

The researchers [5] used a FS strategy to build an ensemble classification model. This model's classification process takes only a few characteristics into account. Ensemble learning, a genetic algorithm, FS, and the results of biomedical tests were used to develop a classification strategy for detecting HD. A conclusion that can be drawn based on these findings is how FS benefits vary depending on the ML approach employed. An accuracy of 97.57 percent was found across all datasets considered for the best-proposed model, which is based on a combination of the genetic algorithm and the ensemble learning model.

In [2], the author explains a method for predicting HD that can quickly and accurately identify abnormalities. In this passage, the method is outlined. Stochastic gradient boosting and a method that removes features recursively are used in the selection process. The Harris-Hawk optimization clustering method is used first to cluster the features, and then the data are classified based on the formed clusters. Next, a more advanced form of a deep genetic algorithm is used to process the classification of the features. An improved genetic algorithm is used to boost the initial weights of a deep neural network, resulting in better overall performance. The neural network's performance improves due to this algorithm's improvements to the weight recommendations it uses.

It stands for a hybrid genetic algorithm (GA) and particle swarm optimization (PSO), which is what GAP SO-RF stands for. The random forest is well-suited to this strategy (RF) [29]. To identify the most accurate predictors of HD, it was created. In the first step of the proposed GAPSO-RF method, a multivariate statistical analysis is used to select the characteristics of the initial population that are most important. Discriminative mutations are carried out after GA is applied. GAPSO-RF uses a GA modified for the global search, and a PSO for the local search. The idea of rehabilitating unsuccessful candidates was also successfully implemented by PSO. The experiments showed that the GAPSO-RF method could accurately predict HD with a 95.6 percent accuracy on the Cleveland dataset and a 91.4 percent accuracy on the Statlog dataset, according to the findings.

Using a genetic algorithm-based FS and an ensemble deep neural network model to predict HD is proposed in [30]. This technique is referred to as a hybrid. A 0.04 learning rate and an Adam optimizer were used in conjunction with this algorithm to enhance the original model. The proposed algorithm's 98 percent accuracy in predicting HD is higher than the accuracy of the methods that preceded it. For example, random forest, logistic regression, SVM, and Decision Tree algorithms have all been shown to be less accurate and faster than the proposed hybrid deep learning-based approach.

Recent research suggests that a hybridized Ant Lion Crowd Search Optimization Genetic Algorithm (ALCSOGA) could be a valuable tool for FS [31]. Three components make up this hybrid optimization: Ant Lion, Crowd Search, and Genetic Algorithm. The ant lion algorithm is used to determine who will be in the elite group. The Crow Search Algorithm, on the other hand, uses each crow's location and memory to analyze the objective function. Combining these two algorithms and then feeding them into the Genetic Algorithm can significantly improve the efficiency of the FS procedure. A stochastic learning rate optimized for long-term memory is used to categorize the extracted optimized features (LSTM). As a result of the testing, the proposed system's ability to predict HD was found to be superior to other, more traditional methods.

Predicting HD may be made easier with a fast and accurate method that can detect abnormalities in a patient's heart rhythm in just a few minutes [2]. Stochastic gradient boosting and a method that removes features recursively are used in the selection process. The Harris-Hawk optimization clustering method is used first to cluster the features, and then the data are classified based on the clusters that were formed. Next, a more advanced form of a deep genetic algorithm is used to process the classification of the features.

An improved genetic algorithm is used to boost the initial weights of a deep neural network, resulting in better overall performance. The neural network's performance improves as a result of this algorithm's improvements to the weight recommendations it uses. When compared to the most advanced techniques currently available, the newly developed method for FS and classification achieves an accuracy of 98.36 percent and 97.36 percent.

Researchers [32] used the Grey-wolf with Firefly algorithm and the Differential Evolution Algorithm to select features for the ANN efficiently, and the Differential Evolution Algorithm was used to fine-tune the hyper parameters of the ANN. Thus, in order to more accurately classify its chosen features, it has been dubbed the Grey Wolf Firefly algorithm with Differential Evolution (GF-DE). Using the proposed classification model, the neural network is trained to obtain optimal weights, and many hyperparameters are effectively tuned. The results of this study demonstrate an efficient and accurate way to predict cardiovascular disease, making it ideal for the task.

The authors of the paper [33] propose a SVM based on differential evolution for the early and accurate detection of heart conditions. Several different data preprocessing techniques are employed before the classification step. One-hot encoding, FS using differential evolution, and normalization are just a few of the techniques available. To ensure that the results are as precise as possible, this is done. This method is evaluated using the Z-Alizadeh Sani and Cleveland datasets, as well as a genetic algorithm-based SVM, which has become increasingly popular in recent years (N2GC-nuSVM). Our experiments show that our differential evolution-based SVM outperforms all of the other evaluated algorithms to arrive at this conclusion. Following an evaluation of the efficacy of various algorithms, this conclusion was reached. The proposed method for diagnosing HD has an accuracy range of 95.1 percent to 86.2 percent when applied to standard datasets. According to the study's findings, HD diagnosis and prognosis could benefit from genetic algorithms – use of the UCI's ML repository's HD data. The ML repository at the UCI, houses the cardiovascular disease dataset. The proposed method uses this dataset. Patients with various heart conditions can benefit from improved classification and prognostication accuracy as a result of this.

A FS model was also used while developing the proposed intelligence system model in [34]. In order to make a diagnosis of coronary artery disease, this model was created. A genetic algorithm and a SVM are two of the tools that are used in the FS process. It is the diagnostic system's job to make decisions, and the system's performance is measured using the precision, sensitivity, positive predictive value, and area under the curve (AUC) criteria described above. The z-Alizadeh sani model FS dataset was used in the tests. Out of the 54 features already present, this dataset generates 5 new features. A performance with an area under the curve of 93.7 percent, accuracy of 87.7 percent, and sensitivity of 87.7 percent can be achieved when these five characteristics are combined.

When the cost of the examination is considered, it is possible to say that the genetic algorithm-based FS model performs reasonably well. The drop in performance is not enough to be considered significant, and the result, which has an AUC of 93.7 percent, is still considered "very good."

2.6 CONCLUSION

Because the conventional approaches to medical history are unreliable for diagnosing HD, the doctor will need to perform a battery of tests in order to arrive at an accurate diagnosis. Diagnostic tests that require invasive procedures take a long time and cost money. If we can accurately diagnose diseases by using more advanced forms of EI in our diagnostic methods, this could have a significant impact on the field of medicine. The results of the most recent study show that cardiovascular disease has been the leading cause of death worldwide for the past twenty years. So, in this day and age, making a quick and accurate diagnosis is vital. This chapter aims to bring attention to some of the problems with using ML techniques to make predictions about HD. This article looks at some of the most important research projects that have used EAs in this field.

REFERENCES

1. Tsao, C. W., Aday, A. W., Almarzooq, Z. I., Alonso, A., Beaton, A. Z., & Bittencourt, M. S. (2022). Heart disease and stroke statistics — 2022 update: A report from the American heart association. *Circulation*, *145*(8), e153–e639.
2. Balamurugan, R., Ratheesh, S., & Venila, Y. M. (2022). Classification of heart disease using adaptive Harris Hawk optimization-based clustering algorithm and enhanced deep genetic algorithm. *Soft Computing 26*, 2357–2373. https://doi.org/10.1007/s00500-021-06536-0
3. Goranova, M., Contreras-Cruz, M. A., Hoyle, A., & Ochoa, G. (2020). "Optimising antibiotic treatments with multi-objective population-based algorithms," *2020 IEEE Congress on Evolutionary Computation (CEC)*, 1–7. https://doi.org/10.1109/CEC48606.2020.9185489
4. Saranya, G., & Pravin, A. (2021). Feature selection techniques for disease diagnosis system: A survey. In Jude Hemanth, D., Vadivu, G., Sangeetha, M., Emilia Balas, V. (eds.). *Artificial intelligence techniques for advanced computing applications* (pp. 249–258). Singapore: Springer.
5. Abdollahi, J., & Nouri-Moghaddam, B. (2022). A hybrid method for heart disease diagnosis utilizing feature selection based ensemble classifier model generation. *Iran Journal of Computer Science*. https://doi.org/10.1007/s42044-022-00104-x

6. El-Rashidy, N., El-Sappagh, S., Islam, S. M. R., El-Bakry, H. M., & Abdelrazek, S. (2021). Mobile Health in remote patient monitoring for chronic diseases: principles, trends, and challenges. *Diagnostics*, *11*(4), 607.
7. Muhammad, L. J., & Algehyne, E. A. (2021). Fuzzy based expert system for diagnosis of coronary artery disease in Nigeria. *Health and Technology*, *11*(2), 319–329.
8. Diwakar, M., Tripathi, A., Joshi, K., Memoria, M., & Singh, P. (2021). Latest trends on heart disease prediction using machine learning and image fusion. *Materials Today: Proceedings*, *37*, 3213–3218.
9. Xu, Y., & Liu, J. (2021). High-speed train fault detection with unsupervised causality-based feature extraction methods. *Advanced Engineering Informatics*, *49*, 101312.
10. Kondababu, A., Siddhartha, V., Kumar, B. B., & Penumutchi, B. (2021). A comparative study on machine learning based heart disease prediction. *Materials Today: Proceedings*.
11. Ashish, L., Kumar, S., & Yeligeti, S. (2021). Ischemic heart disease detection using support vector machine and extreme gradient boosting method. *Materials Today: Proceedings*.
12. Khan, A. H., Hussain, M., & Malik, M. K. (2021). Cardiac disorder classification by electrocardiogram sensing using deep neural network. *Complexity*, *2021*.
13. Mukherjee, R., Sadhu, S., & Kundu, A. (2022). Heart disease detection using feature selection based KNN classifier. In Gupta, D., Polkowski, Z., Khanna, A., Bhattacharyya, S., Castillo, O. (eds.). *Proceedings of data analytics and management* (pp. 577–585). Singapore: Springer.
14. Mienye, I. D., & Sun, Y. (2021). Improved heart disease prediction using particle swarm optimization based stacked sparse autoencoder. *Electronics*, *10*(19), 2347.
15. Ghosh, P., Azam, S., Karim, A., Jonkman, M., & Hasan, M. Z. (2021, May). Use of efficient machine learning techniques in the identification of patients with heart diseases. In Gokhale, A. (ed). *2021 the 5th International Conference on Information System and Data Mining* (pp. 14–20). Silicon Valley, CA: Illinois State University.
16. Premsmith, J., & Ketmaneechairat, H. (2021). A predictive model for heart disease detection using data mining techniques. *Journal of Advances in Information Technology*, *12*(1), 14–20.
17. Kaur, J., & Khehra, B. S. (2021). Fuzzy logic and hybrid based approaches for the risk of heart disease detection: state-of-the-art review. *Journal of the Institution of Engineers (India): Series B*, 1–17.
18. Shorewala, V. (2021). Early detection of coronary heart disease using ensemble techniques. *Informatics in Medicine Unlocked*, *26*, 100655.
19. Bharti, R., Khamparia, A., Shabaz, M., Dhiman, G., Pande, S., & Singh, P. (2021). Prediction of heart disease using a combination of machine learning and deep learning. *Computational Intelligence and Neuroscience*, *2021*. https://doi.org/10.1155/2021/8387680
20. Alarsan, F. I., & Younes, M. (2019). Analysis and classification of heart diseases using heartbeat features and machine learning algorithms. *Journal of Big Data*, *6*(1), 1–15.
21. Ghiasi, M. M., Zendehboudi, S., & Mohsenipour, A. A. (2020). Decision tree-based diagnosis of coronary artery disease: CART model. *Computer Methods and Programs in Biomedicine*, *192*, 105400.

22. Nagavelli, U., Samanta, D., & Chakraborty, P. (2022). Machine learning technology-based heart disease detection models. *Journal of Healthcare Engineering, 2022.* https://doi.org/10.1155/2022/7351061

23. Ko, H., Lee, S., Park, Y., & Choi, A. (2022). A survey of recommendation systems: Recommendation models, techniques, and application fields. *Electronics, 11*(1), 141.

24. Brites, I. S., Silva, L. M., Barbosa, J. L., Rigo, S. J., Correia, S. D., & Leithardt, V. R. (2022, February). Machine learning and IoT applied to cardiovascular diseases identification through heart sounds: A literature review. In Rocha, Á., Ferrás, C., Méndez Porras, A., Jimenez Delgado, E. (eds.). *International conference on information technology & systems* (pp. 356–388). Cham, Switzerland: Springer.

25. Magesh, G., & Swarnalatha, P. (2021). Optimal feature selection through a cluster-based DT learning (CDTL) in heart disease prediction. *Evolutionary Intelligence, 14*(2), 583–593.

26. Tawhid, M. A., & Ibrahim, A. M. (2020). Hybrid binary particle swarm optimization and flower pollination algorithm based on rough set approach for feature selection problem. In *Nature-inspired com.*

27. Koppu, S., Maddikunta, P. K. R., & Srivastava, G. (2020). Deep learning disease prediction model for use with intelligent robots. *Computers and Electrical Engineering, 87*, 106765.

28. Kavitha, D., & Ravikumar, S. (2021). IOT and context-aware learning-based optimal neural network model for real-time health monitoring. *Transactions on Emerging Telecommunications Technologies, 32*(1), e4132.

29. El-Shafiey, M. G., Hagag, A., El-Dahshan, E.-S. A., & Ismail M. A. (2022). A hybrid GA and PSO optimized approach for heart-disease prediction based on random forest. *Multimedia Tools and Applications, 81*, 18155–18179. https://doi.org/10.1007/s11042-022-12425-x

30. Verma, K., Bartwal, A. S., & Thapliyal, M. P. (2021). A genetic algorithm based hybrid deep learning approach for heart disease prediction. *Journal of Mountain Research, 16*(3), 179–187.

31. Kalaivani, K., Uma Maheswari, N., & Venkatesh, R. (2022). Heart disease diagnosis using optimized features of hybridized ALCSOGA algorithm and LSTM classifier. *Network: Computation in Neural Systems, 33*(1–2), 95–123. https://doi.org/10.1080/0954898X.2022.2061062

32. Deepika, D., & Balaji, N. (2022). Effective heart disease prediction with grey-wolf with firefly algorithm-differential evolution (GF-DE) for feature selection and weighted ANN classification. *Computer Methods in Biomechanics and Biomedical Engineering, 25*(12), 1409–1427.

33. Idrees, A., Gilani, S. A. M., & Younas, I. (2022). Automatic prediction of coronary artery disease using differential evolution-based support vector machine. *Journal of Intelligent and Fuzzy Systems: Applications in Engineering and Technology, 43*(4), 5023–5034.

34. Lutimath, N. M., Ramachandra, H. V., Raghav, S., & Sharma, N. (2022). Prediction of heart disease using genetic algorithm. In Gupta, D., Khanna, A., Kansal, V., Fortino, G., Hassanien, A. E. (eds.). *Proceedings of second doctoral symposium on computational intelligence. Advances in intelligent systems and computing,* vol 1374. Singapore: Springer,. https://doi.org/10.1007/978-981-16-3346-1_4

Diabetes Prediction and Classification

3

3.1 INTRODUCTION

Diabetes is a significant health concern in Western countries, and its prevalence rapidly increasing in other parts of the world. It is not a disease but rather a group of diseases that share symptoms, signs, and complications but have unique causes. It is not a disease in and of itself; instead, it is a collection of symptoms. These conditions are now referred to as diabetes mellitus type 1 (T1DM), diabetes mellitus type 2 (T2DM), "other" diabetes mellitus, and gestational diabetes mellitus (GDM) [1]. A wide variety of complications fall under the "other" category; however, we will not be discussing them here because they only affect less than one percent of people with diabetes. GDM is a condition that affects about 5–6 percent of pregnant women and, in most cases, is a precursor to type 2 diabetes. T1DM affects about 5 to 10 percent of people who have diabetes, whereas T2DM is responsible for the condition in the remaining 90 percent of people with diabetes [2]. Insufficient insulin is the underlying cause of all diseases that comprise diabetes syndrome. This deficiency is something that all people with diabetes share in common with one another. Insulin resistance is a problem for people with type 2 diabetes because it makes insulin less effective. This is a problem because insulin resistance makes insulin less effective. Insulin deficiency is a complicated issue that may have several different causes, not all of which are fully understood at this point.

DOI: 10.1201/9781003254874-3

3.2 DIABETES TYPES

The term "overweight metabolic syndrome" (OM) is a catchall term for a group of diseases that cause high blood sugar levels to persist for an extended period. Different manifestations of the same disease can be referred to as diabetes subtypes [3]. They are broken down into categories based on the various entry points through which people can contract the disease.

T1DM, also known as insulin-dependent diabetes or diabetes that develops in young people, is a form of diabetes caused by an autoimmune disease that leads to the destruction of beta cells in the pancreas. This damage is brought about by macrophages and activated C04+ and C08+ T cells that have made their way into the pancreatic islets. Adults between the ages of 20 and 40 years are typically diagnosed with type 2 diabetes for the first time. Evidence suggests that a person's genes and environment each play a role in determining how likely they are to develop this form of diabetes. The human leukocyte antigen (HLA) gene, located on chromosome 6, has been found to have a significant association with type 1 diabetes through research into genetics. The regular cells found within the body, as well as infectious and non-infectious agents found in the outside world, can be distinguished from one another with the assistance of the HLA proteins found on the cell's surface. The autoimmune response that is a sign of type 1 diabetes is triggered when there is a defect in the HLA proteins. The conclusion of this reaction can be found in the cells. The diabetic retinopathy (DR) gene, located close to the HLA gene, has been linked to type 1 diabetes. According to several pieces of evidence, certain viruses might be to blame for type 1 diabetes. [Requires further research] Idiopathic diabetes is a subtype of T1 0M that does not involve autoimmunity but can occur in some patients. This form of diabetes can occur in patients who have T1 0M. Although it occurs less frequently than the autoimmune form of TI OM, it can also affect people of African and Asian descent. Because the individual lacks antibodies that target beta cells, they cannot produce insulin and have an increased risk of developing ketoacidosis [1]. The origins of the disease as well as its progression are not completely understood.

3.3 TYPE 2 DIABETES MELLITUS

Insulin resistance is the primary contributor to type 1 diabetes, which results from the body's inability to produce or secrete an adequate amount of insulin. Around the world, approximately 90 percent of people diagnosed with diabetes have type 2, making it the most prevalent form of the disease. Most of the time, a diagnosis is made in a person's forties. As people get older, their chances of developing type 2 diabetes and the number of people who already have the condition increase. Obesity-related diabetes and non-obesity-related diabetes are the two subtypes that can be further subdivided under the umbrella of type 2 diabetes. It is common for people who are obese and have type 2 diabetes to develop resistance to the insulin that is produced by their bodies because cell receptors change. This is related to how fat is distributed throughout the stomach area. Insulin resistance at the post-receptor level is present in individuals with type 2 diabetes who do not have an excessive body fat. In addition to this, there is a decrease in the production and release of insulin. The dramatic shifts in diet and lifestyle that have taken place in developing countries are a significant contributor to the high prevalence of diabetes in these regions [4]. T2DM, also known as adult-onset diabetes, is brought on by metabolic issues and being overweight.

3.4 GESTATIONAL DIABETES

GDM is a form of diabetes that only manifests itself in a pregnant woman and disappears after the birth of her child. After giving birth, a woman's glucose levels typically return to normal. As a result of the significant changes that occur in women's blood sugar levels during pregnancy and the gestational period, these individuals frequently experience feelings of hunger much more rapidly than is typical for them. In conjunction with the fact that the placenta produces more insulin and the baby's body gradually becomes less sensitive to insulin by the end of the first trimester, the baby develops a temporary insulin resistance. Even though this form of diabetes disappears after pregnancy, it can still lead to several complications, some of which may persist indefinitely. For instance, GDM is a potential teratogen that significantly increases the likelihood that both the mother and the fetus will pass away while the mother is carrying her child. Also, the development of diabetic nephropathy (DN) in GDM can lead to preeclampsia, which is linked to a number of problems with the development of the fetus, including intrauterine growth retardation (IUGR), premature delivery, and stillbirth [5].

3.5 THE DIFFERENT DIABETES TYPES

3.5.1 Retinopathy and Associated Disorders

Even though DR is a common and distinct microvascular complication of diabetes, it is still the leading preventable cause of blindness in people of working age. It is present in one-third of diabetics and has been linked to an increased risk of life-threatening systemic vascular complications such as stroke, coronary heart disease, and heart failure. The best way to reduce the likelihood of retinopathy developing or becoming more severe is to maintain tight control of blood glucose levels, blood pressure, and possibly even blood lipids [6]. The most common cause of DR is microangiopathy, which manifests as changes in the precapillary arterioles, capillaries, and venules of the retina. The damage is due to microvascular leakages, which are caused by a break in the inner blood-retinal barrier, and microvascular occlusions, which are also responsible for the break. The use of fluorescein angiography makes it possible to differentiate between these two diseases so that they can each be treated in their unique manner.

A microaneurysm *is* a small sac-like pouch that can be caused by a localized stretching of the walls of the capillaries. This type of stretching can lead to microaneurysms. They typically take the form of a few red dots close to the macula and are frequently the first sign of retinopathy that a medical professional will notice in a patient. Most retinal hemorrhages appear as "dots" or "blotches" in the dense middle layers of the retina. In extremely unusual instances, the top layer of nerve fibers can experience bleeding in the form of a flame. The majority of the time, extremely high blood pressure is to blame for this kind of bleeding. Yellow lipid deposits with precise edges make up hard exudates characterized by their consistency. They frequently manifest themselves at the periphery of microvascular leakage and have the potential to form a circinate pattern around a microaneurysm that is leaking. They may join forces to form enormous sheets of slime. Edema of the retina is caused by microvascular leakage, and its presence indicates that the barrier between the blood and the retina has been compromised. It can be a challenging to see because of hard exudates in the macula. It gives the thickening of the retina the appearance of spots that are greyish. Because of the thickening, the macula has a higher risk of developing a cyst that resembles a flower petal, which can result in a significant reduction in one's field of vision [7].

DR in its early stages is relatively mild and does not spread. This is the beginning stage of DR, which can be identified by the presence of a few enlarged blood vessels in specific areas of the retina. Medical professionals give these areas of growth the name microaneurysms. At this stage, if even trace amounts of fluid are allowed to seep into the retina, the macula will swell, which will result in microaneurysms. This section of the retina is located relatively close to the optical center of the eye. Moderate DR that has not yet progressed to proliferative stages is the hallmark of stage 2 diabetic eye disease.

As the tiny blood vessels grow larger, they obstruct the flow of blood to the retina, which can cause damage to the retina. This indicates that the retina does not receive an adequate supply of nutrients. As a consequence of this, blood and various other fluids begin to accumulate in the macula. This stage of DR is referred to as severe nonproliferative diabetic retinopathy (NPDR), the third stage of DR.

A significant reduction in the amount of blood that can reach this region is due to the blockage of a more significant number of the blood vessels found in the retina. When this occurs, the body receives messages instructing it to begin the formation of new blood vessels in the retina. These messages are sent by the retina.

The most advanced stage of diabetes-related eye disease is called proliferative diabetic retinopathy (PDR). It is the final stage of DR, which is also the fourth stage. During this stage, the retina will continue to develop new blood vessels. Fluid can more easily escape from these blood vessels, which is because these blood vessels tend to be weak. In the most severe cases, this can result in vision problems such as blurriness, a reduced field of vision, and even blindness.

In order to determine what the characteristics of DR are, researchers use a variety of methodological approaches. The step of image preprocessing is a significant one that contributes significantly to the enhancement of feature extraction and classification. This is since it enhances the characteristics and qualities of the raw fundus images, making it more straightforward for an intelligent system to comprehend the significance of what they depict. It evens out the intensity of the structures, improves the image's contrast, reduces lighting errors and blurriness, eliminates noise, and finds small patterns (subtle lesions) that would otherwise be hidden by structures with a high level of intensity. It makes it possible for an intelligent system to find DR in its development's early and middle stages. This demonstrates how critical it is to clean up artifacts and get rid of backgrounds.

Several different methods, including contrast enhancement contrast stretching morphological operations [8], histogram thresholding and histogram equalization [9, 10], smoothing [11], post-processing [12], shade-correction [13], and illumination equalization/correction [14], are utilized to determine required features.

There are many different techniques for segmentation [15], such as region growing, thresholding, bottom-hat transform, unsupervised segmentation techniques, watershed transformation, active contour model, image reconstruction, template matching, and ensemble-based techniques.

3.5.2 Renal Pathology Nephropathy

Diabetic nephropathy, also known as DN, is recognized as one of the primary contributors to end-stage renal disease in developed and developing nations worldwide. A DN is a form of chronic kidney disease (CKD) that can be caused by diabetes [16]. The kidneys are responsible for regulating the amount of water and salt that are present in the body. This is essential for maintaining healthy blood pressure and guarding against heart and blood vessels damage.

When a person has diabetes, whether it is type 1, type 2, or gestational diabetes, their body cannot produce or use insulin effectively. This is true regardless of which type of diabetes they have. Diabetes mellitus, which occurs during pregnancy and is also referred to as "gestational diabetes," has been associated with an increased risk of developing type 2 diabetes in the years to come.

Diabetes can be diagnosed when there is an abnormally high sugar level in the blood. When left unchecked, high glucose levels in the body can cause damage to organs like the kidneys and the heart over time. DN is a condition in which the kidneys become damaged as a result of diabetes.

3.5.2.1 Stages

By examining the glomerular filtration rate (GFR) [17], also referred to as the percentage of functional kidney tissue, a physician can determine the stage of kidney disease a patient is experiencing.

Stage 1: The kidneys have been damaged, but they still function normally, and the GFR is at least 90 percent.

Stage 2: The kidneys have been damaged, there has been some loss of kidney function, and the patient's GFR is between 60 and 89 percent.

Stage 3: Moderate to a severe loss of kidney function and a GFR that falls between 30 and 59 percent.

Stage 4: A GFR between 15 and 29 percent, along with a significant loss of function.

Kidney failure is indicated by a GFR of less than 15 percent and is present at this stage.

DN is a condition in which the kidneys become damaged due to diabetes. The most common complication of diabetes, DN, is also the leading cause of death among people with diabetes. Numerous researchers have turned to statistical methods to discover the factors that put diabetic patients at an increased risk of passing away. A significant amount of investigation has also been put into the study of the factors that can be utilized in the diagnosis of DN. You will likely develop renal disease, other microvascular lesions, and macrovascular disease if you have had high blood sugar and high blood pressure for a significant amount of time. Patients with type 1 diabetes or type 2 diabetes who took part in clinical trials have repeatedly shown that lowering their glycosylated hemoglobin (HbA1C) levels is linked to a lower risk of developing clinical and structural signs of DN. This finding has been demonstrated over and over again. Logistic regression analysis was selected as the methodology for the study so that prediction rules could be developed to identify diabetic patients who are at a high risk of complications and to investigate risk factors. Both a causal probabilistic network and a time series method were utilized in the research. In addition, the report recommended that diabetic patients use telemedicine and telecare services. Others made use of the collected data in conjunction with several artificial neural networks (ANNs) or decision trees to determine risk factors and forecast when diabetes would first manifest itself [18–20]. The majority of these papers, on the other hand, attempted to forecast when the onset of diabetes would occur, even though complications of the disease are far more significant for both the length and quality of life. Even though many people have tried to solve health issues by utilizing data mining techniques, there are a lot of challenges and restrictions associated with this. The most challenging aspect is compiling all of the information. It is complicated to obtain a dataset that is large enough for this kind of investigation because hospitals have not used electronic medical record systems in a very long time. Additionally, there have been instances where physicians and other medical professionals have resigned from their positions at university hospitals to take jobs at other medical facilities. This could result in different care plans being developed for each patient, including various approaches to conducting interviews and physical examinations. Some outpatients do not visit the hospital as frequently as they should because they are overconfident regarding their state of health or for other personal reasons. Because of all of these factors, the data gathered in a clinical setting might not be very reliable, they might not be sufficient, or they might not even exist. Because of this, it is difficult to obtain useful information from the data that is currently available. The algorithm is the root of a wide variety of additional challenging issues in medical data mining. Learning algorithms such as decision trees, ANNs, and support vector machines (SVMs) have been utilized in several studies in an effort to determine whether or not a patient has a disease or will develop one in the near or distant future (SVMs) [21]. It is difficult for users to understand what the results mean based on the variables that they input into these algorithms, although these algorithms are very good at generalization. This is especially true for ANNs and SVMs, which is why these algorithms are sometimes referred to as "black box" algorithms. Even though logistic regression (LR) is not very good at generalization, it is still widely used as the gold standard by doctors and other medical professionals. It provides the odds ratio for each input variable in addition to the percentage of variation found in the output variable, also referred to as the probabilistic output. In the realm of information retrieval, there are a lot of different approaches to select features (also called variables). These methods are used to rank the importance of the input features, and the end goal is to improve the performance of generalization by removing features that are a barrier to progress. Most of these methods are primarily concerned with how the features that are brought in should be ranked, as opposed to concentrating on what each feature tells us. However, in a real clinical setting, it may be useful for doctors to understand how each variable affects the results. This is an explanation of how the prediction result would change if the value of a feature were changed.

3.5.3 Neuropathy

Diabetes puts a person at risk of developing nerve damage. This damage, known as neuropathy, can cause a great deal of discomfort. It can occur for a variety of reasons; however, it appears that they are all connected to having blood sugar levels that are abnormally high for an extended period. Controlling the blood sugar level should be a priority for everyone, and meet the primary care physician to avoid developing this condition. Diabetes can cause several types of neuropathies, including peripheral, autonomic, proximal, and focal neuropathies, which you and your doctor can discuss.

3.5.3.1 Neuropathy of the extremities

Nerves that are located outside of the brain and spinal cord can suffer damage from the condition known as peripheral neuropathy. The hands and feet of the vast majority of people who have this form of neuropathy are affected by pain, numbness, and weakness. It is also possible for it to have an effect on the body's digestion, urination, and blood flow, among other organs and functions. The brain and spinal cord are both components of what is known as the central nervous system. The peripheral nervous system handles the information that is processed in the brain and sent out to the rest of the body. In addition, the nerves in the body's extremities are responsible for transmitting sensory information to the central nervous system. Peripheral neuropathy can be brought on by a wide variety of conditions and circumstances, including those that are inherited, infections, metabolic issues, traumatic injuries, and toxic exposure. The condition known as diabetes mellitus is one of the most typical causes of this. The majority of patients with peripheral neuropathy report that the pain feels like it is stabbing, burning, or tingling. The majority of the time, symptoms improve, mainly if the condition that was the source of the symptoms can be treated. Certain medications can help ease the discomfort caused by peripheral neuropathy.

Autonomic neuropathy is a disorder of the nervous system. Injuries to the nerves responsible for the body's involuntary functions can lead to a condition known as autonomic neuropathy, a disease. Sexual function, blood pressure, the ability to regulate the temperature in the body, digestion, and even bladder function could be impacted [22].

The damage to the nerves makes it difficult for the brain to communicate with the other autonomic nervous system components, including the heart, blood vessels, and sweat glands. Infections and other health problems, in addition to diabetes, can also cause autonomic neuropathy; however, diabetes is by far the most common cause of this condition. If you take certain medications, there is a possibility that you will suffer nerve damage as a side effect. Because different nerves are affected by the condition, each patient's symptoms and treatment will be unique.

3.6 EA FOR DIABETES

Applying the principles of evolutionary intelligence in the quest to find a cure for DR. In order to gain an understanding of how DR develops and the symptoms it causes, it is essential to be familiar with the signs of occlusion and leakage in the retina. The disease DR is microangiopathy that causes blood vessels to leak and close off [23]. Capillaropathy, changes in the blood, and blockage of small blood vessels are the three components that make up the pathogenesis of DR. When someone has diabetes, what changes do you typically see in their blood vessels? A high blood sugar level is responsible for several changes in the blood vessels. These changes include capillaropathy, in which the walls of the blood vessels become damaged; hematological changes, in which the blood cells change shape, and the blood becomes thicker; and microvascular occlusion, which results in problems with blood flow and lower levels of oxygen.

The following are signs that someone has DR:

- Manifestation symptomatic of DR.
- Microaneurysms.

This is one of the indicators that NPDR is present. It is a locally-based outpouching of the capillary wall that spreads, becomes thicker, and moves outward in a particular region. Plasma can leak from these microaneurysms into the retina because the blood-retinal barrier has been broken or thrombosed. Additionally, tiny microaneurysms stick out of the tiny blood vessels due to the broken blood-retinal barrier. These minuscule red dots will first manifest themselves close to the fovea as the first sign of DR.

On the other hand, it is difficult to spot them when looking at the fundus. You will easily see them during the fundus fluorescein and geography examinations. Near the fovea, a cluster of tiny red dots is the initial manifestation of the disease. It is an image of an eye that has previously been examined with a fundus fluorescein angiogram; the microaneurysms visible in this image are depicted as tiny dots. Microaneurysms developed in the retina due to changes in the blood vessels in that tissue. It has the appearance of a circle and is marked with a few dark red spots, now known as the macula.

3.6.1 Hemorrhages

They are easily confused with dot hemorrhages, which are more noticeable on the retina because they are larger and look very similar to them. This makes it easy to make a mistake. This is the case because they are comparable. Dot-blot hemorrhages may be larger versions of microaneurysms, which appear to be the same on the retina [24]. Hereditary spots and retinal hemorrhages can present themselves in the retinal nerve fiber layer, where they appear as flames, or in the middle layers of the retina, where they appear as red dots. Both of these conditions can impair vision. "Bleeding" is the term used to describe the leaking of blood vessels. It is important to note that although they appear to be red spots, their borders are not identical, and their densities vary.

3.6.2 Hard Exudates

Hard exudates form in the area where the normal retina meets the swollen retina in a patient who has retinal edema. They have a waxy appearance, a yellow color, and precise edges. They are not only composed of macrophages that are loaded with lipids but also lipoproteins themselves. They are almost always found in clusters or rings around the retina and frequently surround microaneurysms in the eye. Because of this configuration, the retina is protected from injury. After the leakage has been stopped, it may take several months or even years for these to disappear [25] completely.

3.6.3 Soft-Consistency Effluents

Hard exudates are characterized by their yellowish appearance, lipid composition, and typical location close to the macula. Their distinct edges can be identified and brought on by bleeding from blood vessels. Spots that look like cotton wool, on the other hand, are made up of axonal debris and are more prevalent in the area close to the optic nerve. This occurs because the cast is deposited in that location by the nerve fiber. Cotton wool spots can be light yellow or white, whereas hard exudates have a darker yellow color. In addition to this, there are a more significant number of cotton-wool spots located close to the optic nerve. Cotton wall spots are non-defined cloud formations that have the appearance of billowy clouds because they lack clear edges. Hard exudates are brought on by blood-vessels leaking, while soft exudates are brought on by blood- vessels blocked. There are cotton wall spots located on the surface of the organs' skin. Most cases of chronic ischemic stroke [26] are brought on by the narrowing or blocking of an artery. Ischemia of the retinal nerve fiber layer (RNFL) occurs when there is a reduction in the amount of blood that flows to the retina. This causes the flow of axoplasm to become more sluggish, and it also causes axoplasmic debris to accumulate along the axons of retinal ganglion cells. This accumulation, more commonly referred to as cotton wool spots (CWS), can be observed in the RNFL as lesions that appear white and fluffy [27]. Numerous mathematical models and a variety of perspectives on DR have been developed and put forward in recent years. Because it has been demonstrated that traditional methods have their limitations, we will be taking into account things like complicated features, a large number of computational complexities, the presence of many solutions, and so on. As a result of evolutionary computing, there will be more than one approach to solving DR and other problems of a similar nature. The concepts of natural phenomena and human centricity will serve as the foundation for these approaches, and they will be used to drive an affinity, effectiveness, and treatment concept throughout society and industry. There will be some investigation into how it can be utilized for DR, and as a consequence, numerous evolutionary computing (EC) algorithms that can be utilized for DR will be described.

People have come to realize that using EC approaches, which are primarily predicated on the concept of naturally occurring bird qualities, is the most effective way to construct a model. The term "evolutionary computing" refers to a computing strategy that is based on the processes that occur in nature. Most of the approaches and models developed to locate and diagnose DR are founded on EC algorithms. This is even though there have been many different strategies and models developed. When more complicated aspects, a variety of potential solutions, and computational challenges are considered, conventional methods become ineffective. As a direct consequence of this, EC and swarm intelligence have recently been applied in an effort to solve these various issues. This article discusses the various responses that EC has given to DR's efforts throughout its history. We investigate how the EC algorithms for the DR processes function in this particular environment to understand better how they operate.

3.7 GENETIC PROGRAMMING

Natural selection is the mechanism that drives evolutionary change in living things. The genetic algorithm (GA), is a method that is based on natural selection. The GA can provide solutions for problems that have constraints as well as problems that have no constraints [9]. GA are a powerful tool for solving problems that involve optimizing designs on a large scale. Researchers are interested in both the mechanics behind how GAs function as well as their applications. The optimization and search algorithm known as GA is a form of meta-heuristic search. It is founded on Darwin's theory that evolution occurs due to natural selection working overtime. The scientific concept of "natural selection," also known as "survival of the fittest," is an essential component of evolutionary thought. The process of natural selection causes organisms to alter their behaviors in order to make it simpler for them to survive in the environments in which they find themselves. The following are a few examples of how GA typically operates:

Step 1: Create a population that is comprised of individuals chosen at random.

Step 2: Determine the level of physical fitness that each individual possesses.

Step 3: The patient gets to choose who will be their parents.

Step 4: Raise a family of children.

Step 5: Evaluate children.

Continue to iterate over Steps 3–5 until either a solution is found. This satisfies the fitness requirements, or a predetermined number of generations have passed.

To develop more effective solutions, If a chromosome has a higher fitness value, there is a greater chance that it will be passed down to the subsequent generation. An estimated fitness value, the probability of a mutation (Pm), and the probability of a cross are used by the GA to determine which offspring to produce (Pc). The process of creating populations will continue indefinitely until all of the criteria have been satisfied and the optimal solution has been identified. Because of its high level of dependability, GA has been put to use in a diverse set of contexts. One of these is known as data retrieval (DR), and it is typically advised to be used to select features and enhance classifiers.

Wei et al. [28] describe a variety of EC techniques for segmentation. Blood vessel segmentation in the automatic DR process can be completed using this feature extraction method. It is a technique for extracting various characteristics. They are attempting to extract from the retina the blood vessels, texture, optic disc, and entropies by combining GA, SVM, and the Bacterial Foraging Algorithm (BFA). The process begins with segmentation, continues with the extraction of image features using bifurcation points, texture, and entropy, and concludes with the extraction of statistical features. After removing the statistical features, the authors then use GA, BFO, and a neural network to separate the images into the three distinct categories of normal, NPDR, and PDR. After that, they look for the most effective way to categorize retinal lesions and assign them mild, moderate, or severe grades.

Kailasam et al. [29] created a very effective computer-aided design (CAD) model by combining GA and Adaptive Neuro-Fuzzy Inference System (ANFIS) in order to classify and find DR abnormalities. This resulted in a very accurate model. It has been demonstrated that this model is accurate. For the most part, the data from optical coherence tomography (OCT) was what Santos et al. [26] looked at in order to determine how well the SVM classifier worked. In order to make this classification, OCT data are the only ones that are used. The method that has been suggested makes use of a variety of instruments and approaches, such as the gathering of data and the application of an OCT data model. It is founded on the logarithmic of linear spaces. After distinguishing between the eyes of diabetic patients and those of healthy people, conduct a heuristic search using a GA. This will help you find the optimal parameter selection. The discrimination process begins once the validation of the parameter of choices for selecting the SVM data model has been completed.

3.8 BLOOD VESSEL DIVISION AND SEGMENTATION

Miere et al. [30] demonstrate an approach to organizing things that preserves the most engaging aspects of the whole. Images of fundus autofluorescence (FAF), which are used to demonstrate segmentation and in which GA plays a significant role, were utilized by the author. However, GA can be perceived differently depending on who is doing the observing and who is experiencing. Images of the fundus's own autofluorescence are what are used to demonstrate segmentation. Classification and segmentation are performed with the assistance of both the GA quantification and this automatic quantification when determining how far along age-related macular degeneration (AMD) is and how to diagnose it. This is done so in order to determine how far along AMD is. They are looking for areas of the vessel structures in the retina where the GA does not fluoresce, and they are finding those areas. Muzammil et al. [31] proposed a segmentation system that could find blood vessels in retinal images by looking at things like dark spots, bright spots, and other things. This system would look for blood vessels in the image by analyzing things like dark spots, bright spots, and other things. The author developed a method to diagnose tractional retinal detachment using images of the retina as the basis for his or her work. When it comes to identifying bright spots from dark ones, this system has the potential to perform extraordinarily well and achieve remarkable results. The acronym RIS, which stands for "retinopathy image segmentation," refers to the system that these people have developed. It correctly divides each of the areas. An approach referred to as "retinopathy image segmentation," which is founded on genetics, was utilized by the author of this study in order to determine which RIS system parameters are the most crucial genetic-based parameter detector–retinopathy image segmentation (GBPD-RIS) [32]. In addition, they developed something known as tractional retinal detachment disease (TRDD. Images of the retinas taken from diabetic patients serve as the foundation for this system. The TRDD comprises three components: the gray-level co-occurrence matrix (GLCM) analysis, the basic frequency model (BFM), and the GBPD-TRDD system.

3.9 CONCLUSION

In the preceding decades, diabetes has been linked to a growing variety of health problems that have been reported worldwide. DR is the most prevalent complication of diabetes. DR has a significant impact on the human retina and results in total blindness. Few essential DR characteristics include macular edema (MA), hard exudates (HE), homogeneous equilibrium (HEM), soft exudates (SE) and new vessels on the disk (NVD). Traditionally, physicians manually examine patients and make societal judgements. It is quite difficult to identify DR with complex characteristics. In this context, in order to combat the current issue of DR, industries are using CAD and machine learning (ML) algorithms. The ML-based CAD system allows for optimal diagnosis decisions and early detection of DR. The automatic DR detection model utilizes DR photos as input to conduct a variety of trials on models derived from distinct algorithms. EA algorithms play a crucial role in the identification of DR at an early stage. The EC and optimization algorithms of GA, particle swarm optimization, ant colony algorithm, bee colony algorithm, and cuckoo search are employed to create an autonomous CAD system for earlier identification of DR. In this chapter, a number of genetic systems and methodologies designed for DR are examined and evaluated based on the approach's strengths and weaknesses. Following preprocessing, feature extraction, segmentation, and classification, the automatic DR models are classified. Methods and applications of salient features for feature extraction and classification are investigated in depth.

REFERENCES

1. Hill-Briggs, F., Adler, N. E., Berkowitz, S. A., Chin, M. H., Gary-Webb, T. L., Navas-Acien, A. ... & Haire-Joshu, D. (2021). Social determinants of health and diabetes: A scientific review. *Diabetes Care*, *44*(1), 258–279.
2. Eckel, R. H., Bornfeldt, K. E., & Goldberg, I. J. (2021). Cardiovascular disease in diabetes, beyond glucose. *Cell Metabolism*, *33*(8), 1519–1545.
3. Rahimi, G. R. M., Yousefabadi, H. A., Niyazi, A., Rahimi, N. M., & Alikhajeh, Y. (2022). Effects of lifestyle intervention on inflammatory markers and waist circumference in overweight/obese adults with metabolic syndrome: A systematic review and meta-analysis of randomized controlled trials. *Biological Research for Nursing*, *24*(1), 94–105.

4. Ekundayo, T. C., Falade, A. O., Igere, B. E., Iwu, C. D., Adewoyin, M. A., Olasehinde, T. A. ... Ijabadeniyi, O. A. (2022). Systematic and meta-analysis of mycobacterium avium subsp. Paratuberculosis related type 1 and type 2 diabetes mellitus. *Scientific Reports*, *12*(1), 1–11.

5. Barani, M., Sargazi, S., Mohammadzadeh, V., Rahdar, A., Pandey, S., Jha, N. K. ... & Thakur, V. K. (2021). Theranostic advances of bionanomaterials against gestational diabetes mellitus: A preliminary review. *Journal of Functional Biomaterials*, *12*(4), 54.

6. Dai, L., Wu, L., Li, H., Cai, C., Wu, Q., Kong, H. ... & Jia, W. (2021). A deep learning system for detecting diabetic retinopathy across the disease spectrum. *Nature Communications*, *12*(1), 1–11.

7. Teo, Z. L., Tham, Y. C., Yu, M., Chee, M. L., Rim, T. H., Cheung, N. ... & Cheng, C. Y. (2021). Global prevalence of diabetic retinopathy and projection of burden through 2045: Systematic review and meta-analysis. *Ophthalmology*, *128*(11), 1580–1591.

8. Hussain, L., Alsolai, H., Hassine, S. B. H., Nour, M. K., Duhayyim, M. A., Hilal, A. M. ... & Rizwanullah, M. (2022). Lung cancer prediction using robust machine learning and image enhancement methods on extracted gray-level co-occurrence matrix features. *Applied Sciences*, *12*(13), 6517.

9. Kaur, G., & Kansal, S. (2021, November). Power law transformation based weighted clipping tri-histogram equalisation. In Gupta, P. K., Gandotra, E., Tyagi, V., Sharma, A. (eds.). *2021 Sixth International Conference on Image Information Processing (ICIIP)* (Vol. 6, pp. 455–460). Himachal Pradesh, India: IEEE.

10. Zhou, J., Zhang, D., & Zhang, W. (2021). Adaptive histogram fusion-based colour restoration and enhancement for underwater images. *International Journal of Security and Networks*, *16*(1), 49–59.

11. Abdulrahman, A. A., Rasheed, M., & Shihab, S. (2021). The analytic of image processing smoothing spaces using wavelet. *Journal of Physics: Conference Series*, *1879*(2), 022118.

12. Li, W., Pan, B., Xia, J., & Duan, Q. (2022). Convolutional neural network-based statistical post-processing of ensemble precipitation forecasts. *Journal of Hydrology*, *605*, 127301.

13. Conde, M. V., McDonagh, S., Maggioni, M., Leonardis, A., & Pérez-Pellitero, E. (2022). Model-based image signal processors via learnable dictionaries. *Proceedings of the AAAI Conference on Artificial Intelligence*, *36*(1), 481–489.

14. Jiang, Y., Quan, X. Q., Xing, Y., Da-yong, L., Xiong, Z., & Sun, Q. (2022). Design of optical imaging system for full-ocean-depth low-light colors. *Optics and Lasers in Engineering*, *154*, 107042.

15. Sarma, R., & Gupta, Y. K. (2021). A comparative study of new and existing segmentation techniques. *IOP Conference Series: Materials Science and Engineering*, *1022*(1), 012027.

16. Miyabe, Y., Karasawa, K., Akiyama, K., Ogura, S., Takabe, T., Sugiura, N. ... & Moriyama, T. (2021). Grading system utilising the total score of Oxford classification for predicting renal prognosis in IgA nephropathy. *Scientific Reports*, *11*(1), 1–8.

17. Murton, M., Goff-Leggett, D., Bobrowska, A., Garcia Sanchez, J. J., James, G., Wittbrodt, E. ... & Tuttle, K. (2021). Burden of chronic kidney disease by KDIGO categories of glomerular filtration rate and albuminuria: A systematic review. *Advances in Therapy*, *38*(1), 180–200.

18. Su, B. (2021). Using metabolic and biochemical indicators to predict diabetic retinopathy by back-propagation artificial neural network. *Diabetes, Metabolic Syndrome and Obesity: Targets and Therapy*, *14*, 4031.

19. Elsharkawy, M., Sharafeldeen, A., Soliman, A., Khalifa, F., Ghazal, M., El-Daydamony, E. ... & El-Baz, A. (2022). A novel computer-aided diagnostic system for early detection of diabetic retinopathy using 3D-OCT higher-order spatial appearance model. *Diagnostics*, *12*(2), 461.

20. Rani, M. J., Shanthi, T., & Kaviya, P. (2021). Diagnosis of diabetic retinopathy from fundal photography. *Annals of the Romanian Society for Cell Biology*, *25*(6), 4526–4538.

21. Abdelsalam, M. M., & Zahran, M. A. (2021). A novel approach of diabetic retinopathy early detection based on multifractal geometry analysis for OCTA macular images using support vector machine. *IEEE Access*, *9*, 22844–22858.

22. Eleftheriadou, A., Williams, S., Nevitt, S., Brown, E., Roylance, R., & Wilding, J. P. (2021). The prevalence of cardiac autonomic neuropathy in prediabetes: a systematic review. *Diabetologia*, *64*(2), 288–303.

23. Mata-Moret, L., Hernández-Bel, L., Hernandez-Garfella, M. L., Castro-Navarro, V., Chiarri-Toumit, C., & Cervera-Taulet, E. (2021). Acute panendothelial retinal leakage in a patient with diabetes mellitus type 1 with a poor metabolic control. *Archivos De La Sociedad Española De Oftalmología (English Edition)*, *96*(11), 598–601.

24. Monferrer-Adsuara, C., Castro-Navarro, V., Gonzalez-Giron, N., Remoli-Sargues, L., Ortiz-Salvador, M., Montero-Hernandez, J. ... & Cervera-Taulet, E. (2022). A case of bilateral unusual retinal hemorrhages in a COVID-19 patient. *European Journal of Ophthalmology*, *32*(2), NP123–NP127.

25. Huang, C., Zong, Y., Ding, Y., Luo, X., Clawson, K., & Peng, Y. (2021). A new deep learning approach for the retinal hard exudates detection based on super-pixel multi-feature extraction and patch-based CNN. *Neurocomputing*, *452*, 521–533.

26. Nazarova, M., Kulikova, S., Piradov, M. A., Limonova, A. S., Dobrynina, L. A., Konovalov, R. N. ... & Nikulin, V. V. (2021). Multimodal assessment of the motor system in patients with chronic ischemic stroke. *Stroke*, *52*(1), 241–249.

27. Huang, L., Wang, C., Wang, W., Wang, Y., & Zhang, R. (2021). The specific pattern of retinal nerve fiber layer thinning in Parkinson's disease: A systematic review and meta-analysis. *Journal of Neurology*, *268*(11), 4023–4032.

28. Wei, J., Zhu, G., Fan, Z., Liu, J., Rong, Y., Mo, J. ... & Chen, X. (2021). Genetic u-net: Automatically designed deep networks for retinal vessel segmentation using a genetic algorithm. *IEEE Transactions on Medical Imaging*, *41*(2), 292–307.

29. Kailasam, M. S., & Thiagarajan, M. (2021). Detection of lung tumor using dual tree complex wavelet transform and co-active adaptive neuro fuzzy inference system classification approach. *International Journal of Imaging Systems and Technology*, *31*(4), 2032–2046.

30. Miere, A., Capuano, V., Kessler, A., Zambrowski, O., Jung, C., Colantuono, D. ...
    & Souied, E. (2021). Deep learning-based classification of retinal atrophy using
    fundus autofluorescence imaging. *Computers in Biology and Medicine, 130,*
    104198.
31. Muzammil, N., Shah, S. A. A., Shahzad, A., Khan, M. A., & Ghoniem, R. M.
    (2022). Multifilters-based unsupervised method for retinal blood vessel segmen-
    tation. *Applied Sciences, 12*(13), 6393.
32. Bhandari, S., Rambola, R., & Kumari, R. (2019). Swarm intelligence and evolu-
    tionary algorithms for diabetic retinopathy detection. In Bhandari, S., Rambola,
    R., Kumari, R., (eds.). *Swarm intelligence and evolutionary algorithms in health-
    care and drug development* (pp. 65–92). New York, NY: Chapman and Hall/CRC.

Degenerative Diseases

4

4.1 INTRODUCTION

A degenerative disease will gradually deteriorate an organ or tissue until it is destroyed or will cause it to weaken over time as it spreads throughout the body. There is a wide variety of degenerative diseases, and a number of them are either directly associated with aging or worsen as people age. There are three types of degenerative diseases: cardiovascular, nervous, and malignant. Neurodegenerative diseases (NDD) of the nervous system occur when nerve cells in the brain or peripheral nervous system gradually lose function and eventually die. This may occur in either the central or peripheral nervous system. There is currently neither a treatment nor a cure for NDDs, nor is it possible to slow down their progression. Alternatively, specific treatments have the potential to alleviate the symptoms of these illnesses, regardless of whether they manifest physically or mentally. In the United States, hypertension, coronary artery disease, and myocardial infarction are the three most prevalent forms of cardiovascular disease. The category of neoplastic diseases encompasses both cancer and other types of tumors. Parkinson's disease (PD) and Alzheimer's disease (AD) are merely two conditions that can affect the nervous system [1].

Numerous factors may contribute to the onset of degenerative diseases. In other instances, the progression of symptoms directly results from an unhealthy lifestyle, including an unhealthy diet and unhealthy food choices. Even though there are treatments for several degenerative diseases, there are still many that are incurable. If a patient finds themselves in this situation, the available treatment options are intended to alleviate their symptoms so they can return to their regular routines.

Some forms of degenerative disease, including cancer, are currently treatable. There is currently no treatment available for Parkinson's and

DOI: 10.1201/9781003254874-4

Alzheimer's, among others. The primary goal of treatment for these conditions is to alleviate the patient's symptoms to the greatest extent possible.

Patients have sometimes prescribed medications such as carbidopa-levodopa to treat the symptoms of PD. However, there is currently no cure for the disease itself. Patients who have undergone this treatment have reported a significant reduction in tremors, which improves their ability to walk and move around.

There is neither a treatment nor a cure for AD. Nevertheless, some medications can alleviate a number of the disease's symptoms. Memantine and cholinesterase inhibitors are the most frequently prescribed medications for patients with AD. The effects of these medications are only temporary, but they are effective at preventing the symptoms from worsening. On the other hand, they are merely obstacles to the progression of the disease. Patients will require as much assistance as they can obtain during this time.

In addition to taking medication, people with Alzheimer's must make significant lifestyle changes to combat the disease's symptoms. Not only must they adjust their eating and exercise habits, but also their living arrangements if they are to achieve optimal health. Home environments for Alzheimer's patients must be uncomplicated and devoid of clutter. Patients must adhere to their doctor's instructions and ensure that their homes contain only the essential furniture, mirrors, and other safety measures.

4.1.1 Neurodegenerative Disease Classification

The condition known as NDD, which wreaks havoc on the nervous system and worsens with time, is more prevalent than you might think. It is not suitable for one's health in any way. The quality of life of patients suffering from NDDs can be improved by analyzing the dynamics of their gait. Researchers have used changes in gait dynamics over time and the size of stride-to-stride fluctuations as a measure of walking physiology to determine the amount of pathological change that occurs in the locomotor control system in healthy people and people with NDDs.

4.1.1.1 Alzheimer's disease

AD is a progressive brain disorder that causes people to lose their short-term memories, become paranoid, and believe things that are not true. Alzheimer's patients often get the wrong diagnosis, thinking these symptoms are caused by stress or aging. AD must be treated with medicines every day. AD is a long-term illness that can last for many years or for the rest of a person's life. Giving the right medication at the right time is essential to prevent serious brain damage. Early detection of this disease is complicated and expensive because it requires collecting much data, using complicated prediction tools, and having an experienced medical professional participate. Since automated systems do not make the same mistakes that people make, they can be used in systems that help doctors make decisions. Images (MRI scans), biomarkers (chemicals, blood flow), and numbers taken from MRI scans have all been used to study AD. This study is based on research that has already been done on AD. So they could tell whether or not a person had dementia. By automating the process of diagnosing AD, not only will it take less time, but there will also be less need for people to talk to each other.

In the early stages of AD, most people can do daily tasks without help. In some situations, the person can still go to work, drive, and do other social things. Even so, the person may still have anxiety or memory problems, like having trouble remembering familiar words or places. Family and close friends of the person have noticed that they cannot remember names as well as they used to. After a thorough medical interview, a doctor may find that a patient has problems with memory and concentration. In the early stages of AD, people often have trouble with:

- It is challenging to think of the right word or name at the right time.
- Having difficulty learning and remembering new people's names when first meeting them.
- Spending each day in a mentally taxing environment, such as a workplace or social setting, can be detrimental to one's mental health.
- They have recently read something but then forgotten it, regardless of whether it was in a book or somewhere else.
- They have been unable to locate a valuable item or misplaced it.

Organizing and planning one's activities and responsibilities are becoming progressively more difficult.

Patients with AD typically experience symptoms for a more extended period as the disease worsens. As the disease progresses, patients with dementia lose the ability to communicate, lose the ability to control their environment, and eventually lose the ability to even move around on their own. They struggle to find the right words and phrases to express themselves when coping with mental illness. People whose memories and cognitive abilities continue to decline may require substantial assistance in their daily lives. At this stage, patients might need assistance with the following:

- Personal care and day-to-day activities 24 hours a day, seven days a week.

- The person loses consciousness of both their immediate surroundings and the most recent experiences that they have had.
- It is becoming more challenging to communicate effectively with other people.
- Infections, particularly pneumonia, are seen more frequently in the general population.

4.1.1.2 Parkinson's disease

Approximately one million people in the United States are affected by PD, the most prevalent form of neurodegenerative movement disorder [2, 3]. In addition to tremor, rigidity, bradykinesia/akinesia, and unstable posture, the clinical picture also includes other motor and non-motor symptoms (NMSs). Although the patient's symptoms are the most critical factor in determining a diagnosis, some tests can help differentiate this parkinsonism from others. Both the death of dopaminergic neurons in the substantia nigra pars compacta (SNpc) and the buildup of misfolded alpha-synuclein in the form of Lewy bodies (LBs) are pathological signs of PD. These pathological signs occur in the SNpc. Once neurodegeneration has been discovered, it will have already spread to other central nervous system (CNS) regions.

In most cases, medical professionals cannot determine what caused a patient's illness; however, in between 5 and 10 percent of these cases, different genetic causes have been identified. Dopamine replacement therapy is the standard treatment for PD, but individuals who have had the disease for extended periods may benefit more from other treatment modalities, such as deep brain stimulation (DBS). No treatment available can stop neurodegeneration, the progression of the disease, or the worsening of a patient's disability at this time. However, the treatments available today are very effective at controlling motor symptoms.

PD can cause tremors, one of the possible symptoms of the disease. Tremor often begins in one of your hands or fingers, but this isn't unheard of. A pill-rolling tremor will occur when the thumb and forefinger are rubbed against one another in a back-and-forth motion. At rest, your hand may shake slightly.

Everything was moving at a snail's pace at the time (bradykinesia). Having PD can slow down the movements over time, making even the most minor tasks difficult and time-consuming. While walking, it is possible to notice that the strides are becoming shorter. Muscles that are tense and muscular muscle stiffness can strike anywhere on the body at anytime. Muscles in the body may begin to ache due to the inability to move normally.

As a result, the posture and balance will change. With PD, the posture may become hunched and lose the center of gravity.

A person's inability to complete a series of tasks without paying attention to each step. The ability to smile, blink or swing the arms while walking may be lost. Different people speak in different ways, and you may not be able to understand what they are saying. The possibility exists that you will speak more slowly and muffle or slur your words during the interview. Your voice has fewer inflections than is typical for someone of your age, making it sound more like a robot than a human.

There are a variety of ways to go about writing. Because of this, writing may become more complex, and your writing may appear cramped. Involuntary movements, emotional instability, and cognitive decline are all symptoms of Huntington's disease, a neurodegenerative disease that affects the brain (cognition).

Huntington's disease affects people between the ages of 30 and 40 years, and a person is most likely to be diagnosed with Huntington's disease. Due to Huntington's disease, which is the most common form of a degenerative brain disorder, Anxiety, depression, small involuntary movements, and poor coordination are all early signs and symptoms that may be present in people with autism. People with Huntington's disease are more likely to experience the involuntary jerks and twitches known as chorea. As the disease progresses, so do the tremors and movements that resemble tremors in intensity. People with this condition may find it more challenging to perform basic activities such as walking, talking, and swallowing. Many sufferers experience changes in their mental health and a decrease in their ability to think and reason. Once symptoms appear, a median life expectancy of 15 to 20 years is predicted for patients with adult-onset Huntington's disease.

There are fewer cases of Huntington's disease in children than in adults. Specifically, it refers to the type of disease that affects children and adolescents. A person's mobility and mental and emotional stability are also negatively impacted by this. Slurred speech and rigidity are also signs of the disease in children. The slow movement, clumsiness, a tendency to trip and fall, and a tendency to bump into things are all symptoms of PD. Drooling is also a possibility in this scenario. Although this is not always the case, it is not uncommon for people's academic performance and reasoning abilities to decline with age. Between 30 and 50 percent of children who have been diagnosed with this condition are affected by it. A typical life expectancy for patients with juvenile-onset Huntington's disease is between 10 and 15 years after the onset of symptoms. In comparison to the adult-onset form, this disease progresses more rapidly.

4.2 EARLY PREDICTION OF NEURODEGENERATIVE DISEASE AND CHALLENGES

4.2.1 Early Prediction – Alzheimer's Disease

AD, a NDD that is unavoidable and invariably fatal, will kill all elderly people. AD is one reason Alzheimer's patients' memories and other mental skills worsen over time. This kind of dementia causes most cases of dementia. AD affects about 5 percent of the people over 65 in developing countries, but it affects 30 percent of people over 85, a very high number. By the year 2050, about 0.64 billion people are expected to have AD. In the last ten years, there have been many attempts to make computer-aided models that can decode MRI images by using machine learning (ML) strategies on the data. These models would get their information from the pictures. ML can help with early diagnosis, analysis of medical images, and the discovery and development of new treatments by using a variety of high-dimensional data sources [4]. During the preclinical stages, ML could accurately tell if a patient had AD or a mild cognitive impairment (MCI). Using ML with multiple high-dimensional data sources can help with early diagnosis, analysis of medical images, and finding and making new treatments [5]. The most crucial difference between Support Vector Machine (SVM) and Artificial neural network (ANN) is whether or not a model can accurately spot Alzheimer's and mild cognitive impairment in the early stages of optimization. This can either convert or not, depending on your taste. SVM gives you a global solution, while ANN gives you an optimal local solution. The better choice is SVM. Both SVM and ANN need to have features taken out [6]. Shi et al. [7] suggested that neural networks and intelligent agents be used together to improve the way medical images are processed. On the other hand, DL uses a learning model that includes feature extraction [8]. DL has been shown to work well with large datasets, especially ones with images. Some researchers used ensemble methods to improve how well they could classify Alzheimer's [9]. In the past few years, many academic studies have looked at how ML could be used to treat Alzheimer's. In 2017, Litjens et al. [10] published a study about how DL techniques can be used to process medical images. Even though the models used in DL are often called "black boxes," statistical methods can be used to figure out how uncertain the network is. Shen and his colleagues studied how DL affects AD [8, 11]. This study suggests that DL models include some uncertainty in their predictions. In 2018, Jose and his colleagues published a study that looked at how well different neuroimaging methods treat neurological disorders. Between 2006 and 2016, Pellegrini et al. [3] published 111 papers about ML algorithms for treating dementia and cognitive impairment. It stressed the importance of using ideas from different fields to make unique ML models. Rathore et al. [12] overviews AD and its early stages. A recent study found that ML algorithms can help make early AD predictions better. The Open Access Series of Imaging Studies (OASIS) 2 neuroimaging dataset is used for this study to create new ML models for unbiased data classification. Through a series of cognitive and neuroimaging tests, dementia is easy to spot. Using ML in these tasks not only makes them better but it also helps make new systems.

4.2.2 Early Prediction – Parkinson's Disease

PD is a NDD affecting dopamine-producing brain cells and the CNS. There is currently no treatment for this condition. A person with Parkinson's must learn to live with the symptoms for the rest of their lives and make accommodations for them. Due to advances in medical science, it is now possible to slow the progression of a patient's condition. We still do not comprehend what causes a person to develop a disease for which there is no cure. Therefore, the most effective way to treat PD at this time is to obtain an accurate diagnosis as soon as the first symptoms appear.

Raundale et al. made the Parkinson's Telemonitoring Vocal Data Set available so they could test the algorithms they had proposed. Maunika and Rao [13] suggested implementing the K-Nearest Neighbor (KNN), State-of-the-Art Supervised ML Algorithm with k equal to 5 to achieve an accuracy of 97.43 percent. In addition, they have shown that deep learning (DL) algorithms accurately predict the severity of PD in patients. The authors [14] investigated PD using three distinct ML techniques: the SVM, the KNN, and the ANN. After evaluating these three procedures, it was determined that they had an accuracy level close to 100 percent for the dataset being evaluated. Typically, medical professionals who work in medical centers, such as doctors or medical representatives, assess the severity of PD. This has been a common practice for a considerable amount of time. Researchers [15] devised an innovative method for diagnosing PD using remote smartphone monitoring of patients suspected of having the disease. In addition to data regarding the user's postural instability, dexterity, gait, and tremor, the smartphone continuously records the user's voice. Utilizing the framework's two-step procedure, the framework's features are chosen. Utilizing all of this information, they were able to obtain very encouraging results, which is very encouraging. The authors [16] concluded that the best way to extract the various parts of the brain from structural MRI data would be to use a two-step procedure involving the statistical analysis and a ML technique. When determining the severity of PD, they were able to get within 93.75 percent of the actual value, which was the closest they could get. In addition to transfer learning, researchers employed two CNN models, namely Inception V3 and ResNet50, to determine whether a person had PD. By utilizing the Inception V3 model, they achieved a 96.67 percent accuracy rate to determine whether a person had PD. The algorithms known as Substancia Nigra (SN) and VGG16 CNN were applied to MRI scans in order to identify significant features. In order to classify PD, the extracted features were combined with three DL algorithms known as ResNet-34, VGG-19, and ResNet-50. Based on the findings of a study that compared and analyzed each method, they determined that ResNet-50 provided the most precise results. The findings of the study led them to this conclusion. On the UCI dataset consisting of 45 acoustic features, Ouhmida et al. [7] achieved the highest achievable level of accuracy, namely 93.10 percent. ANN and convolution neural networks (CNN) were used to complete this task successfully

4.3 EA FOR TREATING DEGENERATIVE DISORDERS

Patients suffering from a wide range of conditions, such as PD, glaucoma, diabetic retinopathy, and breast cancer, can be diagnosed and treated with the help of clinical datasets. In healthcare applications, data mining techniques are frequently used to better understand a patient's medical history through recognizing patterns. This is accomplished using computer programs. The result is an improvement in the care provided to patients. The decisions you end up making due to [data mining] could not be better informed. Because patients' clinical data contains several kinds of uncertainty, it can be challenging to arrive at a decision that is both all-encompassing and certain when working in practice. Computer-aided diagnosis (CAD) systems will significantly assist clinicians in seeking a second opinion before making a final decision and formulating a treatment plan in the case of PD. Several different conceptual frameworks for extracting knowledge from clinical datasets have been put forward. Numerous published works have utilized a wide variety of diagnostic approaches in order to make predictions regarding diseases such as lung cancer [5]. Additional information regarding a diagnostic framework that is comparable to that used for PD can be found in [4].

Every method used for diagnosing degenerative diseases allocates a sizeable portion of its resources to the extraction and classification of features. In the beginning, a supervised learning algorithm needs to be utilized to produce a training set for the classifier. After that, a test set is utilized to assess the classifier's effectiveness. The decision trees (DT), the random forests (RF), the KNN, and the SVM are some of these SVM. Features largely determine the performance of the classifier in the training set that is both irrelevant and redundant. In order to achieve an increase in performance, it is necessary to get rid of unnecessary features and select the feature subsets that offer the greatest benefit. Various feature selection algorithms [9–11] have been proposed in the published research that has been done in the scientific community.

It is possible that the diagnosis and treatment of degenerative diseases could be aided by applying bio-inspired computational algorithms. The genetic algorithm (GA), particle swarm, firefly bat, bacterial foraging, and flower pollination are some examples of bio-inspired algorithms that were researched in [12]. Even though these models currently hold the lion's share of the market share, additional research is required to improve the efficacy or performance of the many disease prediction models currently in use. A wide variety of ML and DL strategies have been used in various applications. These methods, despite the fact that they boost performance, have difficulty optimizing their parameters and suffer from overfitting.

4.3.1 Genetic Algorithms in Diagnosing Degenerative Disorders (DD)

It has been demonstrated that GAs, which are based on biology and are known as GAs, are constructive in finding solutions to optimization problems involving large search areas. These methods of searching do not have the problems that many other methods, such as complete, greedy, heuristic, and random, do. These problems include being stuck in an optimal local solution and/or having high computational costs. Exhaustive search, heuristic search, greedy search, and random search are some other categories of search strategies. Random search is another type of search strategy. In order to enhance the overall quality of the feature set, many different ML applications, such as classification, clustering, and time-series prediction, have used the feature selection technique. The classification of NDDs and the prediction of their time series are of particular interest. Using time-series classification and time-series prediction, it is possible to classify those who have suffered brain damage and forecast the progression of the disease. It is absolutely necessary

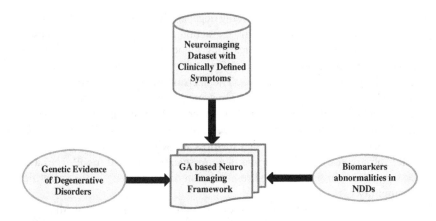

FIGURE 4.1 Genetic algorithm strategy for NDDs.

for you to make frequent use of both of these applications. In order to better describe and diagnose DD, GA has been utilized in the process. The classification of DD has been accomplished with these methods to a level of accuracy that is considered satisfactory (between 70 and 95 percent).

Figure 4.1 shows the GA strategy for identifying NDDs. The GA strategy and the adaptive neuro-fuzzy inference system were put together to make GANFIS, a hybrid diagnostic tool. This tool aims to be more accurate in identifying NDDs. The experiment used benchmark datasets from two NDDs, AD and PD. This was done to prove that the model presented was correct. This was done so that the model would be correct. Several different metrics, such as accuracy, precision, recall, f-score, and kappa coefficient, have been used to evaluate the proposed hybrid approach. Based on the experiments, GANFIS did a much better job diagnosing dementia and PD than the traditional neuro-fuzzy method. The study's authors came up with an idea for a segmentation method that is based on GAs and could be used to accurately and quickly diagnose PD. Complex CNN models would have to be used for this method. Based on the results of the experiments, the proposed method of MRI to classify PD is better than the standard of care that is used now.

Early diagnosis of PD is important for starting treatment, which helps the doctor heal the patient, stop the disease from spreading to other brain cells, and, in the end, save many lives. As a result, this research aims to make a dynamic expert diagnostic system that can accurately predict PD. This system proposes a hybrid method that combines a piece of two-stage mutual information and an auto encoder-based dimensionality reduction approach with a genetically optimized Light GBM (MI-AE-GOLGBM) algorithm to improve the performance of the proposed system and predict the best outcomes. This mixed-method aims to make the proposed system's predictions more accurate. The proposed method, MI-AE-GOLGBM, uses four methods: mutual information, auto encoder, GA, and the LightGBM algorithm. Mutual information and auto encoder are used in this method to create a two-step strategy for reducing the number of dimensions and choosing the most informative features from the input dataset. This method is used to pick out the most useful parts of the dataset. The Light GBM algorithm sorts PD patients from healthy controls by using the newly generated features and the best-optimized hyper parameters from the two-stage mutual information and auto encoder-based dimension reduction methods as well as the GA. This makes the proposed system more accurate and reliable. The Light GBM algorithm uses the GA to optimize the algorithm's hyper parameters smartly. The Light GBM algorithm uses these newly generated characteristics along with the best-optimized hyper parameters that have been given.

Researchers in [6] made algorithms for the three most important parts of diagnosing diseases that get worse over time. These issues are how to tell the difference between people with AD or frontotemporal dementia (FTD) and healthy controls (HC), how to tell the difference between behavioral FTD (bvFTD) and AD, and how to tell the difference between different types of primary progressive aphasia (PPA). Researchers in [7] also made algorithms to tell the difference between AD and FTD (When it comes to diagnosing degenerative diseases, each of these factors is very important. After being changed, KNN and Bayes Net Naive were added to the fitness function of GAs. Compared to principal component analysis, the following is true of these GAs: (PCA). The algorithms used to tell the difference between AD and HC have been checked by someone from the outside. The results back up the use of fluorodeoxyglucose–positron emission tomography (FDG-PET) imaging, which made it possible to diagnose AD, FTD, and other diseases with a very high level of accuracy. GAs were used to figure out which characteristics are most important, and as few features as possible were used in the analysis. These traits may be taken into account when FDG-PET images of the brain are evaluated automatically.

Several studies have been proposed to find and tell the difference between different types of DD. However, most of these studies have only focused on telling the difference between healthy people and people with Alzheimer's. Even so, many studies have been suggested to find and tell the difference between different DD groups. These studies did not find the most reliable biomarkers, which would have led to more accurate results. They also did not use the best hyperparameters, which would have led to the best results possible. So, these studies did not come up with as good results as they could have. The authors of the research paper [8] came up with a new stacking-based ensemble learning system as a way to solve these problems. This system uses the GA tuning method for hyperparameters and the four traditional classifiers already there. When the model was being evaluated, its accuracy, precision, recall, and F1 score were all looked at. The simulation shows that stacking-based ensemble learning, which uses a GA, can tell the difference between the CN group, the MCI group, and the AD group with an accuracy of 96.7 percent, a recall rate of 96.5 percent, a precision rate of 97.9 percent, and an F1-score of 97.1 percent.

In, a new way of processing patient voices and using GAs to choose the neural network architecture for PD patients' monitoring system were created. This was done at the same time as the study. The process of pre-processing a person's voice is explained here so that the main parameters that can be used to evaluate a person's condition can be found more easily. This is done so that the doctor can better understand how the patient is doing right now. It has been shown that how the data is organized directly affects how well GAs work when they are used to build neural networks. To do this, the level of success with different data structures was compared. Because of this, a hybrid approach, in which one part of the neural network architecture is chosen by hand after careful analysis and the other part is built automatically, can give the most accurate results when figuring out how sick a patient is. This is because the hybrid approach takes the best parts of both the old and new ways of doing things. This is because the hybrid approach combines parts of both the old and new ways of doing things.

In this work, the idea is put forward that voice analysis could be used to create a new algorithm for diagnosing PD and that this algorithm could be used. In the first step of the process, a GA is used to choose the best-extracted features. The next step is to use a network based on SVM to sort people into two groups: healthy and those with PD (SVM). This study's dataset is made up of many different biomedical voice signals from 31 people, 23 of whom had PD and 8 of whom were healthy. The results showed that classification accuracy of 94.50 percent could be reached with only four optimized features, 93.66 percent could be reached with only seven optimized features, and 94.22 percent could be reached with only nine optimized features.

The authors evaluated and tested the method to see if it could be used to find early-stage PD. The method is based on using evolutionary algorithms (EAs) to measure bradykinesia in finger tapping (FT) (PD). The study found that an instrument that could accurately measure the severity levels of bradykinesia in people with PD, was easy to use, used classifiers based on EAs, and could tell the difference between normality and early-stage PD could be used to measure the severity levels of bradykinesia (PD) accurately. With the help of a new CAD system, we can accurately predict how mild cognitive impairment (MCI) will turn into AD one to three years before a clinical diagnosis is made. This system uses a GA and a ranking of features to look at data from structural magnetic resonance imaging. With this system, we can predict whether mild cognitive impairment will turn into a disease or not. They showed that combining feature ranking and GAs is an excellent way to accurately predict the progression from MCI to AD and to diagnose AD earlier.

Professionals in the medical field have suggested that GA, wavelet kernels (WK), and Extreme Learning Machines (ELM) could be used as parts of a method for a correct diagnosis of PD. This research uses the ELM learning method to teach the single-layer neural network (SLNN) classifier how to work. The authors of this study were able to figure out the best values for each of these parameters and the total number of neurons that are hidden in ELM by using a GA. Statistical tools like classification accuracy, sensitivity, and specificity analysis, and receiver operating characteristic (ROC) curves are used to figure out how well the proposed GA-WK-ELM method works. The usefulness of these methods depends on how well they work. The GA-WK-ELM method was found to have the highest classification accuracy, which was calculated to be 96.81 percent. This was found out after it was tried out.

The input dataset's statistical distribution dramatically affects how well Convolutional Neural Networks (CNN) diagnose diseases that get worse over time. Depending on the dataset, these results may look very different. Several different hyperparameters in how CNN models are set up can significantly affect how well they converge. Because we have so much data, one of the essential parts of this research is figuring out which parameters should be used to describe the network's structure. The GA, has been getting much attention lately because it is becoming increasingly popular. Its goal is to choose a high-performance network architecture automatically. The authors showed that GA could be used to improve the structure of a network. The search space for GA includes both the configuration of the network structure and the hyperparameters. The authors showed that it is possible to improve how a network is built.

4.4 CONCLUSION

Degenerative disease is a condition that progresses predictably over time. Neurological conditions that last a long time are called degenerative diseases. By 2050, 33 degenerative disease diagnoses per minute are expected. Rapid diagnosis is crucial. Cancer causes up to 70 percent of dementia cases. Unable to recall recent events is a warning sign. Disorientation, mood swings, fatigue, and inability to care for basic needs may worsen as the disease progresses. Isolating themselves from family and friends helps sick people cope with their illness. If a person cannot perform any bodily functions, they will die. Even after a thorough exam, it is difficult to predict a person's lifespan, but the average is 3 to 9 years. This chapter explains how to use the evolutionary approach to extract valuable data from a dataset of degenerative diseases and use that data in ML algorithms to improve prediction accuracy, such as when diagnosing AD. This chapter shows how to use the evolutionary approach to extract valuable data from degenerative disease datasets. The chapter focused on using the evolutionary approach to extract valuable data from degenerative disease datasets. Comparing diseases helped. Early ML accuracy rates ranged from 72 to 84 percent. We used EAs to improve accuracy by 5–10 percent. Our goal was accomplished.

REFERENCES

1. Segato, A., Marzullo, A., Calimeri, F., & De Momi, E. (2020). Artificial intelligence for brain diseases: A systematic review. *APL Bioengineering, 4*(4), 041503.
2. Trinh, J., & Farrer, M. (2013). Advances in the genetics of Parkinson disease. *Nature Reviews Neurology, 9*(8), 445–454.
3. Gusella, J. F., Lee, J. M., & MacDonald, M. E. (2021). Huntington's disease: Nearly four decades of human molecular genetics. *Human Molecular Genetics, 30*(R2), R254–R263.
4. Sudharsan, M., & Thailambal, G. (2021). Alzheimer's disease prediction using machine learning techniques and principal component analysis (PCA). *Materials Today: Proceedings*.
5. Kour, H., Manhas, J., & Sharma, V. (2022). Hybrid system based on genetic algorithm and neuro-fuzzy approach for neurodegenerative disease forecasting. In Gupta, D., Polkowski, Z., Khanna, A., Bhattacharyya, S., Castillo, O. (eds.). *Proceedings of data analytics and management. Lecture notes on data engineering and communications technologies*, vol 90. Singapore: Springer. https://doi.org/10.1007/978-981-16-6289-8_27
6. Sreelakshmi, S., & Mathew, R. (2022). A hybrid approach for classifying Parkinson's disease from brain MRI. In Ullah, A., Anwar, S., Rocha, Á., Gill, S. (eds.). *Proceedings of international conference on information technology and applications. Lecture notes in networks and systems*, vol 350. Singapore: Springer. https://doi.org/10.1007/978-981-16-7618-5_15
7. Dhar, J. (2022). An adaptive intelligent diagnostic system to predict early stage of Parkinson's disease using two-stage dimension reduction with genetically optimized LightGBM algorithm. *Neural Computing and Applications, 34*, 4567–4593. https://doi.org/10.1007/s00521-021-06612-4

8. Díaz-Álvarez, J., Matias-Guiu, J. A., Cabrera-Martín, M. N., Pytel, V., Segovia-Ríos, I., García-Gutiérrez, F., ... & Alzheimer's Disease Neuroimaging Initiative. (2022). Genetic algorithms for optimized diagnosis of Alzheimer's disease and frontotemporal dementia using fluorodeoxyglucose positron emission tomography imaging. *Frontiers in Aging Neuroscience*, 983.

9. Khoei, T. T., Catherine Labuhn, M., Caleb, T. D., Chen Hu, W., & Kaabouch, N. (2021). A stacking-based ensemble learning model with genetic algorithm for detecting early stages of Alzheimer's disease, *2021 IEEE International Conference on Electro Information Technology (EIT)*, pp. 215–222. https://doi.org/10.1109/EIT51626.2021.9491904

10. Shichkina, Y., Irishina, Y., Stanevich, E., & de Jesus Plasencia Salgueiro, A. (2021). Application of genetic algorithms for the selection of neural network architecture in the monitoring system for patients with Parkinson's disease. *Applied Sciences*, *11*(12), 5470. https://doi.org/10.3390/app11125470

11. Kuresan, H., & Samiappan, D. (2022). Genetic algorithm and principal components analysis in speech-based Parkinson's early diagnosis studies. *International Journal of Nonlinear Analysis and Applications*, *13*(1), 591–602.

12. Gao, C., Smith, S., Lones, M.. Jamieson, S., Alty, J., Cosgrove, J. ... & Chen, S. (2018). Objective assessment of Bradykinesia in Parkinson's disease using evolutionary algorithms: Clinical validation. *Translational Neurodegeneration 7*, 18. https://doi.org/10.1186/s40035-018-0124-x

13. Glass, D. J., & Arnold, S. E. (2012). Some evolutionary perspectives on Alzheimer's disease pathogenesis and pathology. *Alzheimer's and Dementia*, *8*(4), 343–351.

14. Beheshti, I., Demirel, H., Matsuda, H., & Alzheimer's Disease Neuroimaging Initiative. (2017). Classification of Alzheimer's disease and prediction of mild cognitive impairment-to-Alzheimer's conversion from structural magnetic resource imaging using feature ranking and a genetic algorithm. *Computers in Biology and Medicine*, *83*, 109–119.

15. Avci, D., & Dogantekin, A. (2016). An expert diagnosis system for Parkinson disease based on genetic algorithm-wavelet kernel-extreme learning machine. *Parkinson's Disease*, *2016*, 5264743. https://doi.org/10.1155/2016/5264743

16. Lee, S., Kim, J., Kang, H., Kang, D. Y., & Park, J. (2021). Genetic algorithm based deep learning neural network structure and hyperparameter optimization. *Applied Sciences*, *11*(2), 744.

Tuberculosis

5

5.1 INTRODUCTION

Tuberculosis, more commonly referred to as TB, is a contagious disease that can be transmitted from one individual to another. It is one of the primary reasons why people pass away all over the world. Before the pandemic caused by the coronavirus (COVID-19), TB was the single disease responsible for the most deaths. The bacillus that causes TB is spread through the air when people with the disease cough or sneeze, such as Mycobacterium TB (e.g., by coughing). Other areas of the body besides the lungs are also susceptible to TB infection. It is statistically more likely for the disease to be found in men than in women (approximately 90 percent of those diagnosed). Approximately one-quarter of the world's population has been identified as having TB. TB is a disease that can be treated, and it is also a condition that can be avoided. The medication course required to treat TB disease typically lasts for a period of six months. Infections caused by TB can be cured with drug regimens ranging from one month to six months. There needs to be universal health coverage, also known as UHC, in place so that anyone who is afflicted with an illness or infection can receive the appropriate medical attention. By addressing risk factors for TB, such as poverty, undernutrition, HIV infection, smoking, and diabetes, it is possible to reduce the number of people who contract the disease and ultimately pass away as a result of it. Because of this policy, there will be a reduction in the total number of people who pass away from TB. However, some countries do not.

The initial actions are that there are fewer than 10 new cases and fewer than 1 death caused by the disease each year for nearly 100,000 people in that country. Scientists will need to make scientific advancements such as the development of a new vaccine to reduce the number of TB cases around the world to the same level they have reached in these countries with a low burden of the disease.

The infectious disease is known as TB poses a significant risk to public health because it can affect people of any age and almost always results in the victim's death. By the year 2021, it is anticipated that 10 million people will be afflicted with the disease, the majority of whom will be women and children (3.3 million women, 5.5 million men, and 1.2 million children) (WHO, 2021). In 2021, a single pathogen infection was responsible for 24 percent of all TB-related fatalities (WHO, 2021). Testing for and correctly diagnosing TB ought to be regarded as the most important and significant advance that can be made [1, 2]. On the other hand, TB infections are notoriously difficult to forecast, and it is not unusual for identification and diagnosis to take a considerable amount of time. A delay in the diagnosis could result in drug resistance, including multidrug resistance (MDR), which occurs when an isolate is resistant to two first-line drugs (rifampicin and isoniazid), and extensive drug resistance (XDR), which includes MDR and also shows resistance to fluoroquinolones and at least one of the injectable drugs. Both of these forms of drug resistance can be prevented by early detection and treatment. When conducting diagnostic tests in the medical field, laboratories frequently make use of high-priced microscopes to examine sputum and other samples from patients. Finding TB patients who are also infected with the disease can be challenging and time-consuming. Because of the similarities in their symptoms, the various forms of TB are typically difficult to differentiate from one another using decision boundaries or distinguishing rules. This is because the diseases share similar symptoms. Because of this, arriving at the appropriate conclusion or diagnosis can be quite challenging. Even though the diseases are difficult to manage, they must be treated as quickly as possible to prevent them from becoming even more severe.

Medical professionals can use machine learning (ML) methods to assist them in making decisions regarding how to diagnose or predict TB. Using a wide variety of statistical and ML techniques, researchers have been able to model and predict the progression of TB disease. Logical regression [3, 4], neural networks [5–8], support vector machines (SVMs) [9, 10], decision trees (DT) [11, 12], naive Bayes [13, 14], and other methods are included in these techniques. Many factors prevent these methods from being used to model actual issues that occur in the real world, even though they make important contributions. The fact that the user is expected to assume a particular model form, which in turn necessitates having an in-depth knowledge of the applicable theory, is the primary drawback of traditional methods. For instance, in order to solve problems involving regression, it is frequently sufficient to locate a set of model coefficients for linear or polynomial functions that accurately describe the input variables. This can be done in a number of different ways. In order to develop these models, you will need to have an understanding of the statistical distribution of the data.

Evolutionary algorithms (EAs) may make it simpler to find solutions to problems that are neither linear nor simple [15]. The term "genetic programming," also abbreviated as "GP," refers to an EA that can be used to find the optimal model in a way that is both original and astute. "GP," in this particular instance, it is not a particular method for resolving issues. Instead, it is a method of developing mathematical models that is more general [16]. GP is superior to other ML methods for modelling or unsupervised learning in situations where it is impossible to predict the form the solution model will take in advance. This is the case in plenty of different scenarios. Through the creation of computer programs of varying lengths, GP has been successful in resolving various medical issues [17].

5.2 TUBERCULOSIS CLASSIFICATION

The TB is an infectious disease almost always caused by the bacterium known as Mycobacterium TB. The majority of the time, these microorganisms are inhaled together with the polluted air that is present. Infections typically travel through the blood and lymph system from the lungs to other body parts. However, the infection can also spread through other organs and the respiratory tract if they are exposed to the virus early. Bacteria can be dispersed through the air when a sick person talks, coughs, sneezes, or spits. A person needs to be exposed to only a trace of bacteria to become ill. The length of time you were exposed to the bacteria and how closely you were to it both play a role in determining your risk of becoming ill. People living or working close to an infected individual have a greater risk of contracting the disease.

People with pulmonary TB are more likely to cough up blood, experience weight loss, lose their appetite, get fevers, sweat at night, and lose their appetite. TB also causes people to lose their appetite. Another symptom of TB is that the disease can spread to other areas of the body, which can also cause the disease. The disease causes the human body to progress through two distinct stages of growth while it is present. It is common practice to refer to this stage as the "TB infection" stage. A progression characterizes the second stage of the disease toward a more authoritarian state [18]. At first, the infection "imports" itself from a host already carrying it. This can only happen when two people come into contact. There are two ways to look at TB: active disease or an infection that has not yet manifested itself. The most common form of active TB is known as pulmonary TB, but the disease can also spread to other organs, which is referred to as "extrapulmonary TB."

5.2.1 Pulmonary

The phrase "about the respiratory system" is how the word "Pulmonary" is described when looked up in the dictionary. When the symptoms of TB become more severe, the organ most commonly affected is the lung. Despite this, the infection has the potential to spread to other parts of the body. Mycobacterium TB is the name of the bacterium that is responsible for the disease known as TB, which is transmitted through the air and destroys body tissue. In most cases, the bacterium known as Mycobacterium TB is to blame for its progression. When Mycobacterium TB primarily infects a person's lungs, that person is said to have pulmonary TB. However, the disease can still move to other parts of the body if it starts here. Early detection and treatment of pulmonary TB can result in a full recovery from the disease.

In North America and Europe during the 18th and 19th centuries, pulmonary TB was commonly referred to as "consumption." This disease was widespread throughout those time periods. Better living conditions and the discovery of antibiotics such as streptomycin and isoniazid allowed medical professionals to treat TB more effectively and put a stop to its further spread.

A trustworthy origination: according to the WHO, TB is one of the leading causes of death worldwide, Developing nations are home to the vast majority of TB victims, both infected and those who succumb to the disease. According to the American Lung Association (ALA), 9.6 million people are living with a form of disease that is still active in their bodies. A citation is required for this. If the disease is not treated, there is a possibility of developing life-threatening complications [3, 4]. Computerized tomography (CT) is currently the method that provides the highest accuracy when detecting TB in the lungs. When TB is suspected in the early stages of the disease, a chest X-ray, also known as a CXR, is typically performed to confirm the diagnosis. This is because CXRs are the most common, require the least amount of radiation, and cost the least amount of money. In addition to this, unexpected pathologic changes may be revealed. Researchers have spent decades trying to develop a computer-aided detection (CAD) system dependent on medical imaging to diagnose TB. To begin, CAD uses algorithms to select and extract useful pathogenic features from images. This provides meaningful quantitative insight into the disease being investigated. Although these methods take a lot of time and rely heavily on the artificial extraction of patterns that contain helpful information, CAD can still provide valuable quantitative insights. Because many diseases are only visible in a small portion of an image at a time, identifying features quickly becomes more difficult as the image is viewed in its entirety. Because of issues such as poor transferability between different datasets and unstable performance in newly generated data, the CAD system can also not make a well-grounded decision with a high degree of accuracy. These issues are caused by a large amount of medical image data and the dynamic nature of the disease, and they contribute to the fact that the CAD system cannot make a decision.

5.2.2 Extrapulmonary

Before the Mycobacterium TB infection can become active, it must first enter a dormant state, which most people already have. Patients are considered to have latent TB as long as they are infected with TB but do not exhibit any symptoms of active TB. When an infection of the lung manifests, it is not uncommon for a dormant infection to suddenly become active and contagious. It can potentially affect the lymph nodes, the central nervous system, the bone and joint system, the genitourinary tract, the abdomen (including the internal abdominal organs and the peritoneum), and the pericardium. The medical term for TB that manifests in areas of the body other than the lungs is called extrapulmonary tuberculosis (EPTB).

According to the Global TB Report published by the WHO, however, additional efforts are required to improve access to diagnosis and treatment, particularly in countries that account for more than half of the global gap. Conventional medical testing can be used to diagnose EPTB in several different ways. Pulmonary TB (PTB)-related research has been conducted a lot more frequently than EPTB-related research has. Due to the paucibacillary nature of EPTB, the disease is challenging to diagnose. A tissue biopsy, Ziehl-Neelsen staining, fluorescent microscopy, and an interferon-release assay are the diagnostic procedures that can be used to diagnose EPTB. In regions with limited resources, there is not uncommon for insufficient laboratory facilities, which precludes the application of these methods. In order to obtain samples from patients, it is necessary to perform invasive procedures [5]. The diagnosis of TB requires several procedures that take a significant amount of time. These procedures include sputum smear tests, cultures, and biopsy specimens.

5.2.3 Challenges in Diagnosing PTB and EPTB

In this part of the chapter, we will examine the various applications of ML and deep learning (DL) that can be used to diagnose TB. It is essential to detect TB at an early stage in order to prevent the disease from spreading.

Even though a correct diagnosis of TB is possible, there are not enough qualified radiologists to go around is a significant obstacle. Diagnostics performed by computers can often produce results that are more accurate and helpful than those performed by humans. It also makes it possible to construct the system with high precision at a cost that is lower than it would be with human diagnostics.

When trying to determine whether or not an individual had TB, researchers utilized a Bayesian Convolution Neural Network rather than a conventional Convolution Neural Network. When the system was put through its paces using the Montgomery and Shenzhen TB datasets, it achieved an accuracy rate of 96.4 percent for the Montgomery dataset and 86.4 percent for the Shenzhen dataset [19]. In order to construct the system, the Montgomery dataset was utilized. In order to detect TB in chest X-rays, researchers came up with the novel idea of using techniques such as segmentation, visualization, and DT classification. The total number of DT models that were utilized for transfer learning was nine (ResNet18 and ResNet50 and ResNet101 and ChexNet and InceptionV3 and Vgg19 and DenseNet201, and SqueezeNet and MobileNet). Cases of TB and other cases won't be confused with one another using these models. U-net models were partitioned the data, and X-ray pictures were analyzed to determine how the data should be categorized. Images of the lungs after being dissected were also used in an experiment at one point. The incorporation of segmented lung regions significantly improved the detection of the system. Through the application of image segmentation, the accuracy increased from 97 percent to 100 percent. According to the researchers' findings, DenseNet201 performed better than segmented lungs, and ChexNet performed better than other DT models [20].

The researchers were required to construct a Deep Convolution Neural Network model to evaluate the model's generalizability. X-rays of the chests of people who suffered from TB were used for this purpose. In order to train the dataset, one group of chest X-rays is used, and another group is used to test the trained dataset. They arrived at the verdict that the diagnostic performance of the model shifts depending on which dataset population is used to train it [21]. This was the conclusion that they came to after conducting their research. Researchers hoped that by utilizing a convolution neural network, they could create a CAD TB system. Additionally, the use of a DT model to locate CXRs is recommended by reference [22]. This study used two different CXR datasets, both of which were available to the general public. One of these datasets came from the National Institutes of Health, and the other came from Shenzhen Hospital. The model was initially fine-tuned with the assistance of the artificial bee colony algorithm, and then the linear average-based ensemble method was utilized to complete the process. It was discovered that the Deep Convolution Neural Network had a better overall performance when the three steps described above were carried out in conjunction with one another [23]. Researchers have proposed a number of potential methods for diagnosing TB. A data-acquisition system can automatically take pictures of the sputum field of view, and the recognition system uses a method called transfer learning (VentNet) [24] to customize itself to the individual patient. When attempting to classify the data using multiclass categorization, the knowledge is typically transferred from one person to another. Because of improvements made to both the specificity and sensitivity of the test, the overall accuracy has increased to 95.05 percent. People have developed CAD systems that make use of image processing in order to locate cases of lung TB. Because of this, it was possible to detect and diagnose diseases more timely and accurately. The proposed system includes components for the cleaning and segmentation of data as well as the extraction of features and classification of data. The findings of this research indicate that it is possible to give the system greater specificity while preserving its existing levels of accuracy and sensitivity. The system's accuracy, sensitivity, and specificity are all above average, coming in at a combined score of 76 percent. When presented with images of sputum microscopy, it has been hypothesized that a deep neural network could identify the location of TB bacilli [3]. The model in question achieves a recall rate of 83.78 percent and a precision rate of 67.5 percent. TB-AI [25] is the name given to a system that was developed by researchers and uses convolution neural networks. This system was developed to identify TB bacilli that can withstand an acidic environment. The training set consists of 45 samples with a total of 30 positive cases and 30 negative cases. The test set consists of 210 samples drawn at random from the population, with 108 positive cases and 93 negative cases. There was a 97.94 percent correlation between how well this system worked and how well it worked. Researchers [26] have devised a method that utilizes digital image processing that involves multiple steps in order to categorize and count ZN-stained TB bacilli. It is used to divide bacilli in the same way that hue color components were previously used. Similarly, the structure of a bacillus is evaluated to determine whether or not it is able to survive in its natural setting. As evidenced by the display of a variety of micrographs, this technique has been demonstrated to be accurate to a level equivalent to 90 percent when counting bacilli. It was suggested that the non-linear SVM could be used as a primary screening method for TB infection [9]. In addition, a gradient EA is used in order to locate the optimal value for each parameter. The classification system does not involve any kind of medical examination and is based on data available to the general public. Another tool used to investigate TB transmission is called a C5.0 DT [27]. It has been discovered that the proposed algorithm produces useful results while simultaneously maintaining a low error rate. Paper 34 presented a novel algorithm for forecasting the finding of TB. This algorithm was used to make predictions. The conventional DT had to be modified in a few key-ways to accommodate this development. Both SVM and random forests (RF) fall short of KNN's 94 percent accuracy when it comes to the detection of TB. The study's findings led the researchers to conclude that this approach, which does not require the participation of technical personnel, is capable of producing satisfactory outcomes.

In order to determine whether or not a person has TB, researchers use the Naive Bayesian method. The data that was utilized in this investigation originated from the medical files of patients who had previously been identified as having PTB or EPTB. The study's findings led the researchers to realize that the Naive Bayesian approach to ML is a viable option for detecting TB. When X-ray and lab results were considered, the correct response was obtained 85.95 percent of the time. Researchers created the most accurate DT model possible with the assistance of the Biomarker Patterns Software (BPS) [18]. When distinguishing between EPTB and non-EPTB, this model has a specificity of 97 percent and a sensitivity of 84 percent. The fact that this model correctly classified data 91.6 percent of the time suggests that the DT model could be helpful in the diagnosis of EPTB.

5.3 EA FOR DIAGNOSING TUBERCULOSIS

Detection of TB typically involves a series of steps beginning with a physical examination and continuing with laboratory testing. The patient's prognosis can be determined using the first method, which consists of interrogating the patient with a series of targeted questions. The second method is used to validate the results of the first method, which was a physical examination. It is possible to skip the diagnosis entirely after a physical examination if the attending physician is confident of the patient's condition, but doing so is not recommended. Treating TB in developing countries can be challenging because many hospitals do not have the necessary medical equipment. Researchers have been looking into how well, and quickly EAs can solve problems that occur in the real world [4], aiming to improve science and medicine. Evaluation of the surrounding environment needs to be incorporated into the diagnostic process for TB to reduce the number of people who get the disease and ultimately pass away from it.

5.3.1 Role of EA in Tuberculosis Treatment

Genetic algorithm-based tuberculosis diagnosis framework is shown in Figure 5.1. The use of GAs, one of a large number of traditional algorithms, has been widespread among researchers working to develop new methods for detecting TB in humans at an earlier stage than has ever been possible. A typical GA assigns a binary array of a fixed length to each individual in the initial population to serve as the genotype for that individual. Calculating the normalized probability of selection is possible by dividing each individual's specific ability by the total ability level. This will allow for the calculation of the normalized probability of selection. The algorithm ultimately chooses the parents of the offspring through the use of a single-point crossover. When analyzing TB, it is necessary to select the initial feature that is most significant in order to prevent the problem of overfitting. The most effective strategy for completing this mission is to use GA.

Stochastic search methods that are founded on natural genetics, such as GAs, have the potential to be very effective in environments that are both

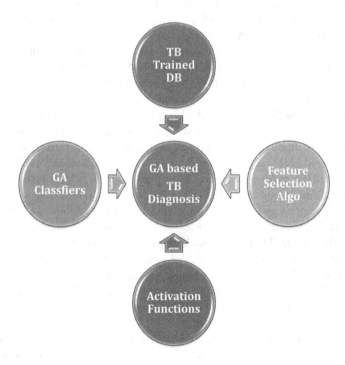

FIGURE 5.1   GA-based TB diagnosis framework.

large and complex. Problems are easier to address and resolve when their presence assists them. It is possible to find a solution to an optimization issue by employing a process known as a GA, which is iterative. A mechanism that encrypts and decrypts information can be used to locate each solution. As a consequence of this, every possible solution needs to be portrayed as a chromosome. This happens over and over again when going in the opposite direction. A feature can be missing or present at the I position; however, a value of zero or one indicates the absence of the feature. GAs always start with a collection of chromosomes chosen at random as their starting point. An evaluation of a solution's efficiency and overall quality can be performed with the assistance of a fitness function. As a result, the fitness function in (1) takes the chromosome as an input and then returns the fitness value that is associated with it. The next thing to do is to figure out which people are in the best physical shape and have the most life experience to bring up the next generation. The creation of new populations from already existing ones can be accomplished by using two different reproductive operators: mutation and cross-breeding.

Subjects who took part in the study [28] The SVM classifier is used in TB drug resistance research and modeling. In order to improve the overall performance of SVM, GA is utilized to select the 20 features from the complete feature set that are most relevant to the problem at hand. In addition, the kernels of the SVM model are analyzed to determine which would be the best fit for the SVM algorithm. We must keep feeding the trained SVM and GA model new data to achieve an accurate prediction of a patient's response to treatment. The SVM combined with GA can accurately predict treatment resistance using data that has not yet been collected, and it only needs 20 features. The findings backed up my assertion in this regard. It has been hypothesized [27] that an improved model that is based on ML could select the appropriate hyper-parameters and extract the most helpful texture characteristics from pictures of TB. For our purposes, it is critical that we simultaneously improve our precision while also reducing the number of extracted characteristics. One of the challenges we face is optimizing our use of multiple tasks at once. The most valuable characteristics for a SVM classifier are selected with the aid of the outcomes of a genetic algorithm (GA) (SVM). We were able to achieve significantly higher accuracy and better results with the method that we proposed by making use of the ImageCLEF 2020 dataset as our primary resource. Compared to the prior art, our results with the proposed method are significantly better.

Auscultation, the most widely used diagnostic method in pulmonology, can differentiate between the many diseases that can affect the lungs and guide the diagnostic process toward more specialized procedures [29]. The creation of a hybrid GANN aims to automate lung sound diagnosis.

An ANN's training parameters and computation time can both be optimized and shortened with the help of this generalized artificial neural network (GANN). The recently developed method is able to distinguish between normal, wheezing, and crackling lung sounds.

Patients who are in a critical condition need to have a measurement taken of their arterial blood partial pressure of carbon dioxide (PaCO2). PetCO2, also known as Pet CO2, is a non-invasive method that can be used to predict PaCO2 in healthy individuals. However, in sicker individuals, it is possible that this method is biased and has less utility. This is the recommended method to use in order to circumvent the challenges that are associated with sampling arterial blood. The authors [30] developed a GA for predicting PaCO2 using capnography data collected from emergency room patients who were not intubated. The proposed system may improve both the accuracy and bias of the PaCO2 prediction.

On the basis of the Genetic-Neuro-Fuzzy Inferential method that was proposed in work [31], medical practitioners can reap the benefits of a decision support platform that assists them in administering a TB diagnosis that is accurate, timely, and cost-effective. [Note: this platform is only available to medical practitioners.] The utilization of this platform may also be of use in the treatment of TB. The performance evaluation found that the sensitivity and accuracy results of 60 percent and 70 percent fell within the acceptable range established by subject matter experts. The case study used 10 patients from St. Francis Catholic Hospital Okpara-In-Land, located in Delta State, Nigeria.

5.4 CONCLUSION

TB detection is critical in medicine because of the alarmingly high mortality rate associated with cancer. As a result, future generations will require tools for diagnosing, treating, and managing TB and its associated diseases. TB research has also incorporated technological methods for early detection and treatment due to global technological advancements and the rise in digitalization. The earlier TB is diagnosed and treated, the lower the number of deaths. Numerous subtypes, stages, and diagnostic challenges are associated with TB, and this chapter provides an overview of them. Aside from that, the chapter discusses the methodology that underpins the most recent diagnostic tools and technologies. Early detection in humans of TB is a significant focus of this paper, as are the various EA approaches that researchers from industry and academia have proposed.

REFERENCES

1. Cheon, S. A., Cho, H. H., Kim, J., Lee, J., Kim, H. J., & Park, T. J. (2016). Recent tuberculosis diagnosis toward the end TB strategy. *Journal of Microbiological Methods*, 123, 51–61.
2. World Health Organization. (2016). *On the road to ending TB: Highlights from the 30 highest TB burden countries. (no. WHO/HTM/TB/2016.06)*. Geneva, Switzerland: World Health Organization.
3. Alsaffar, M., Alshammari, G., Alshammari, A., Aljaloud, S., Almurayziq, T. S., Hamad, A. A. ... & Belay, A. (2021). Detection of tuberculosis disease using image processing technique. *Mobile Information Systems*, 2021. https://doi.org/10.1155/2021/7424836
4. Ejeh, S. O., Alabi, O. O., Ogungbola, O. O., Olatunde, O. O., & Dere, Z. O. (2022). A comparison of multinomial logistic regression and artificial neural network classification techniques applied to TB/HIV data. *American Journal of Epidemiology, 6*, 14–18.
5. Lin, L., She, C., Chen, Y., Guo, Z., & Zeng, X. (2022). TB-NET: A two-branch neural network for direction of arrival estimation under model imperfections. *Electronics, 11*(2), 220.
6. An, L., Peng, K., Yang, X., Huang, P., Luo, Y., Feng, P. ... & Wei, B. (2022). E-TBNet: Light deep neural network for automatic detection of tuberculosis with X-ray DR imaging. *Sensors, 22*(3), 821.
7. Pathak, K. C., Kundaram, S. S., Sarvaiya, J. N., & Darji, A. D. (2022). Diagnosis and analysis of tuberculosis disease using simple neural network and deep learning approach for chest X-ray images. In Mehta, M., Fournier-Viger, P., Patel, M., Lin, J. CW. (eds.). *Tracking and preventing diseases with artificial intelligence* (pp. 77–102). Cham, Switzerland: Springer.
8. Han, D., He, T., Yu, Y., Guo, Y., Chen, Y., Duan, H. ... & Yu, N. (2022). Diagnosis of active pulmonary tuberculosis and community acquired pneumonia using convolution neural network based on transfer learning. *Academic Radiology, 29*(10), 1486–1492.
9. Zhang, Y. D., Wang, W., Zhang, X., & Wang, S. H. (2022). Secondary pulmonary tuberculosis recognition by 4-direction varying-distance GLCM and fuzzy SVM. *Mobile Networks and Applications*, 1–14.
10. Yang, F., Yu, H., Kantipudi, K., Karki, M., Kassim, Y. M., Rosenthal, A. ... & Jaeger, S. (2022). Differentiating between drug-sensitive and drug-resistant tuberculosis with machine learning for clinical and radiological features. *Quantitative Imaging in Medicine and Surgery, 12*(1), 675.
11. Deelder, W., Napier, G., Campino, S., Palla, L., Phelan, J., & Clark, T. G. (2022). A modified decision tree approach to improve the prediction and mutation discovery for drug resistance in mycobacterium tuberculosis. *BMC Genomics, 23*(1), 1–7.
12. Veselsky, A., Pavel, G., Ivan, B., Belik, V., & Lavrova, A. (2021, November). Assessment of bronchial blocking success in patients with destructive forms of tuberculosis using decision trees. In *2021 6th International Conference on Intelligent Informatics and Biomedical Sciences (ICIIBMS)* (Vol. 6, pp. 135–136). Oita, Japan: IEEE.

13. Geetha Pavani, P., Biswal, B., Sairam, M. V. S., & Bala Subrahmanyam, N. (2021). A semantic contour based segmentation of lungs from chest X-rays for the classification of tuberculosis using Naïve Bayes classifier. *International Journal of Imaging Systems and Technology, 31*(4), 2189–2203.
14. Halliru, A., Wajiga, G. M., Malgwi, Y. M., & Maidabara, A. H. (2022). A model for prediction of drug resistant tuberculosis using data mining technique. *Computer Science & IT Research Journal, 3*(1), 1–22.
15. Raghul, S., & Jeyakumar, G. (2022). Parallel and distributed computing approaches for evolutionary algorithms—A review. *Soft Computing: Theories and Applications*, 433–445.
16. Alsalibi, B., Mirjalili, S., Abualigah, L., & Gandomi, A. H. (2022). A comprehensive survey on the recent variants and applications of membrane-inspired evolutionary algorithms. *Archives of Computational Methods in Engineering*, 1–17.
17. Karmakar, R. (2022). Application of genetic algorithm (GA) in medical science: A review. In Luhach, Kumar. A. (eds.). *Second international conference on sustainable technologies for computational intelligence* (pp. 83–94). Singapore: Springer.
18. Migliori, G. B., Tiberi, S., Zumla, A., Petersen, E., Chakaya, J. M., Wejse, C. ... & Zellweger, J. P. (2020). MDR/XDR-TB management of patients and contacts: Challenges facing the new decade. The 2020 clinical update by the global tuberculosis network. *International Journal of Infectious Diseases, 92*, S15–S25.
19. Tasci, E., Uluturk, C., & Ugur, A. (2021). A voting-based ensemble deep learning method focusing on image augmentation and preprocessing variations for tuberculosis detection. *Neural Computing and Applications, 33*(22), 15541–15555.
20. Wong, A., Lee, J. R. H., Rahmat-Khah, H., Sabri, A., Alaref, A., & Liu, H. (2022). TB-Net: A tailored, self-attention deep convolutional neural network design for detection of tuberculosis cases from chest X-ray images. *Frontiers in Artificial Intelligence, 5*, 1–15. https://doi.org/10.3389/frai.2022.827299
21. Mahbub, M. K., Biswas, M., Gaur, L., Alenezi, F., & Santosh, K. C. (2022). Deep features to detect pulmonary abnormalities in chest x-rays due to infectious diseaseX: covid-19, pneumonia, and tuberculosis. *Information Sciences, 592*, 389–401.
22. Zaidi, S. Z. Y., Akram, M. U., Jameel, A., & Alghamdi, N. S. (2022). A deep learning approach for the classification of TB from NIH CXR dataset. *IET Image Processing, 16*(3), 787–796.
23. Nafisah, S. I., & Muhammad, G. (2022). Tuberculosis detection in chest radiograph using convolutional neural network architecture and explainable artificial intelligence. *Neural Computing and Applications, Special Issue*, 1–21.
24. Sudarshan, V. K., Ramachandra, A., Tan, R., Ojha, N. S. M., & Tan, S. (2022). VEntNet: Hybrid deep convolutional neural network model for automated multiclass categorization of chest X-rays. *International Journal of Imaging Systems and Technology, 32*(3), 778–797.
25. Ravin, N., Saha, S., Schweitzer, A., Elahi, A., Dako, F., Mollura, D. ... & Chapman, D. (2022). Mitigating domain shift in AI-based TB screening with unsupervised domain adaptation. *IEEE Access, 10*, 45997–46013.
26. Shetty, D., & Vyas, D. (2022). Combination method for the diagnosis of tuberculous lymphadenitis in high burden settings. *Surgical and Experimental Pathology, 5*(1), 1–7.

27. Sharma, A., Machado, E., Lima, K. V. B., Suffys, P. N., & Conceição, E. C. (2022). Tuberculosis drug resistance profiling based on machine learning: A literature review. *Brazilian Journal of Infectious Diseases*, *26*. https://doi.org/10.1016/j.bjid.2022.102332

28. Kanesamoorthy, K., & Dissanayake, M. B. (2021). Prediction of treatment failure of tuberculosis using support vector machine with genetic algorithm. *International Journal of Mycobacteriology*, *10*(3), 279.

29. Petmezas, G., Cheimariotis, G. A., Stefanopoulos, L., Rocha, B., Paiva, R. P., Katsaggelos, A. K. ... & Maglaveras, N. (2022). Automated lung sound classification using a hybrid CNN-LSTM network and focal loss function. *Sensors*, *22*(3), 1232.

30. Alba, G. A., Samokhin, A. O., Wang, R. S., Wertheim, B. M., Haley, K. J., Padera, R. F. ... & Maron, B. A. (2022). Pulmonary endothelial NEDD9 and the prothrombotic pathophenotype of acute respiratory distress syndrome due to SARS-CoV-2 infection. *Pulmonary Circulation*, *12*(2), e12071.

31. Siddiqui, A. K., & Garg, V. K. (2022). An analysis of adaptable intelligent models for pulmonary tuberculosis detection and classification. *SN Computer Science*, *3*(1), 1–8.

Muscular Dystrophy

6

6.1 INTRODUCTION

Muscular dystrophies (MDs) are a collection of inherited conditions that can be passed down from one generation to the next. Those afflicted with these conditions experience a slow deterioration in their muscular strength in addition to an increase in the severity of their disability. The medical condition known as MD is progressive, which means that it gets worse with time [1]. It is common for it to begin by affecting a specific group of muscles, and then it will spread to affect all of the muscles. This can happen at any point during the progression of the condition. When some forms of MD reach a particular stage in their progression, they can cause damage to the heart and the muscles that control breathing. This contributes to the elevated mortality risk that is associated with the condition. The MD may be able to accomplish this at some point. Even though there is currently no cure for multiple sclerosis, there are treatments that can help alleviate many of the symptoms associated with the disease.

6.1.1 Causes of Muscular Dystrophy

MD is caused by alterations, or mutations, in the genes responsible for controlling the structure and function of a person's muscles. It becomes more difficult for the muscles to perform their typical functions as a direct result of the mutations, which are the factors responsible for the changes in the structure of the muscle fibers. This causes a gradual increase in the severity of the impairment over time. Within families, mutations are likely to be transmitted from one generation to the next. If your family has a history of multiple sclerosis, your primary care physician may advise you to undergo genetic testing and counseling. If this is the case, you should adhere to their recommendations.

DOI: 10.1201/9781003254874-6

This information will allow your physician to calculate the likelihood of you developing multiple sclerosis or having a child with the condition.

6.1.2 Types of Muscular Dystrophy

There are many different kinds of subtypes of MD, and the symptoms of each are different. There are many subtypes, and depending on which one you have, your life expectancy or level of disability will be very different [2]. Some of the most common types of medical diagnoses are as follows:

- A myotonic dystrophy is a form of MD that can affect individuals at any stage. People who do not have myotonic dystrophy have a lower life expectancy than those who do, particularly if they have a severe condition. However, this is not something that occurs every time. Facioscapulohumeral MD can start in childhood or later in life. Because of how slowly it spreads, it is rare to see it in children. The most common and severe form of MD is called Duchenne (DMD). This form of disease affects a large number of people. The majority of patients with this type are male.
- The symptoms of Becker MD appear later in childhood than those of DMD, and the severity of Becker MD is not as severe as that of DMD. In most cases, a person's life expectancy is unaffected by having this condition. The first symptoms of limb-girdle MD, a group of conditions, typically show up in a person in their late teens or early 20s. While some forms of the disease can rapidly deteriorate and ultimately be fatal, other manifestations of the illness progress more slowly but eventually prove fatal. The symptoms of oculopharyngeal MD, a form of MD, typically do not appear in a person until they are between the ages of 50 and 60, and the condition does not typically have an impact on a person's life expectancy. The disease known as Emery-Dreifuss MD affects a significant number of people.

6.1.3 Diagnosing Muscular Dystrophy

The doctor may employ any number of diagnostic procedures in the search for underlying causes of muscle weakness [3]. Blood samples are taken for testing. It is these enzymes that are measured after an injury has occurred when a muscle is injured. Your child's doctor will apply prickly protrusions to various parts of his or her body in order to simulate an electrode placement. After that, your child's doctor will instruct them to contract and release their muscles gradually.

Electrodes are used to determine a subject's electrical activity. Electrodes are inserted into the muscle and connected to a machine. Using a needle, your child will have a small piece of muscle removed. This procedure does not cause any discomfort. They'll use a microscope to see if there are any missing or broken proteins in the sample. This test may reveal what type of MD your child has.

The subject's reflexes, as well as their muscular strength and coordination, will be analysed. Other problems with a patient's nervous system can be detected using these tests, which are helpful to doctors. EKG stands for electrocardiogram. The electrical signals sent from the heart can be used to determine the rate of your child's heartbeat and whether or not it is in a healthy rhythm. Using imaging techniques, it is possible to see how well your child's muscles are working. A magnetic resonance imaging exam, known as an MRI, is an option for them. Magnets and radio waves are used to create images of the subject's internal organs. Sound waves are used in ultrasound imaging to produce images of the internal organs.

Doctors can search for the genes that cause MD using a sample of the patient's blood. In addition to helping doctors figure out what is wrong, genetic testing is a must-have for anyone considering starting a family or with a history of the disease.

6.1.4 Treating Muscular Dystrophy

Currently, no treatment can reverse the effects of MD, but there are many things that can help with the physical difficulties and restrictions that come along with it. Something like [4] might be the case.

The goal of physical therapy is to keep muscles strong and flexible by performing a variety of exercises and stretches on a regular basis.

Through occupational therapy, your child will learn how to use their muscles in the most efficient way possible. In addition to providing therapy to their patients, therapists can also instruct patients on correctly using assistive devices such as wheelchairs, braces, and other similar devices.

People who have difficulty speaking because of weak muscles in their throat or face may benefit from speech therapy to improve their ability to communicate with others.

Respiratory therapy may be an option if your child is having difficulty breathing. They will either learn how to breathe more easily or buy equipment to help them do so. Some of the symptoms may be alleviated by taking medication. DMD can be treated with a variety of medications, including eteplirsen (Exondys 51), golodirsen (Vyondys 53), and vitolarsen (Viltepso). People with a specific mutation in the gene responsible for DMD can receive injections of these medicines as a form of treatment. The use of these medications causes the body to produce more dystrophin, to be more precise. Have a discussion with your child's doctor about any possible side effects.

Medications that can help treat epilepsy and stop muscle spasms.

Patients with heart issues may benefit from taking medication to lower their blood pressure. Medications are known as immunosuppressants slow down the immune system's ability to damage muscle cells.

Your child's breathing may improve, and muscle damage may be slowed if steroids like prednisone and deflazacort (Emflaza) are given early enough. They have the potential to cause serious harm, including bone fractures and an increased risk of illness.

Creatine is a naturally occurring molecule found in every body cell, including the brain and the heart. It is well known to help muscles get more energy and help people get more robust. Before giving your child any dietary supplements, make an appointment with your child's primary care physician.

Some complications associated with MDs, such as heart problems and swallowing difficulties, may be treated surgically. Researchers are always using clinical trials to look for new possible treatments for MD. During these clinical tests, new drugs are looked at to see if they can treat patients and be given safely.

6.1.5 Common Muscular Dsytrophy

When it comes to muscle disease, the most common type of dystrophy is DMD [5]. Men are more likely to get the disease, despite it being possible for females to carry it and experience only mild symptoms. The progression of the disease is characterized by symptoms such as frequent falls, trouble getting out of bed, difficulty running and jumping, a waddling gait, and large calf muscles. Impairments in cognitive development and a slowed growth rate are also early indicators of the disease. Cerebral palsy is associated with an increased risk of falling, a common symptom.

Like Becker's MD, DMD presents milder symptoms and progresses more slowly than Becker. Even in your 20s or 30s, you may not notice any symptoms until you are older. A chance exists for this to happen.

6.2 EARLY CLINICAL DIAGNOSIS OF MD

Imaging is an essential component of the diagnostic process for muscle diseases because it gives the clinician additional information about the presence of the disease as well as the disease's severity, extent, and activity.

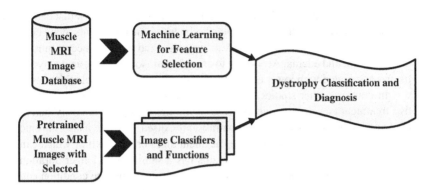

FIGURE 6.1   ML frameworks for muscular dystrophy diagnosis.

Figure 6.1 shows the machine learning (ML) frameworks for muscular dystrophy diagnosis. The doctor may be unable to use the imaging modality that is most effective for diagnosing myopathies: MRI if you have a pacemaker or implant. Using muscle ultrasound to diagnose neuromuscular diseases has become increasingly common since it's easy to use, lack of contraindications, and improved resolution for soft tissue structures have made it a valuable tool. Ultrasound technology advancements are to blame for this development. In myopathies, such as MD, an increase in echogenicity can be easily seen as an increase in connective tissue and fatty replacement. This has been linked to a decline in health-related quality of life and the progression of the illness. Ultrasound, on the other hand, is susceptible to biases introduced by the operator as well as the interpreter. Ultrasound's application is restricted to a more specialized market due to the technology's reliance on changes in echo intensity. This makes comparing ultrasound results from different systems difficult. Some of the problems that have been identified could be solved using quantitative ultrasound and backscatter analysis.

Myositis is an immune-mediated inflammatory muscle disease that has been studied extensively. Dermatomyositis (DM) and polymyositis (PM) are frequently effectively treated and cured by medication. Both conditions affect the proximal muscles, but only diabetes also affects the skin [6, 7]. Inclusion body myositis, also known as IBM, is a difficult-to-treat condition that primarily affects the quadriceps and extremity muscles. This condition results in severe muscle atrophy and fat replacement due to the inefficacy of conventional treatments. Myositis can affect the muscles and the soft tissues surrounding them. The muscles and tissues surrounding the affected area may develop edema, fatty infiltration, subcutaneous necrosis, and even calcification as a direct result [8, 9]. This group faces a difficult challenge when correctly identifying and quantifying the pathology that can be seen on ultrasound in the various stages and forms of the disease. This is especially true regarding changes that can be reversed by treatment. It appears that echo intensity measurements alone cannot distinguish between the disease's active phase and the presence of muscle inflammation and edema. According to an early ultrasound study, acute myositis was associated with lower echo intensities and increased muscle thickness.

In contrast, other studies on juvenile DM have demonstrated that effective treatment causes an acute increase in the muscle's echo intensity before allowing it to return to normal. In contrast to the disease's acute phase, chronic myositis is characterized by fatty replacement and fibrosis, both of which can cause excruciating pain. Chronic myositis has a significantly longer duration than acute myositis. Studies that IBM carried out show that the condition can be effectively differentiated from the disease when screening for it in muscles that are afflicted, such as the flexor digitorum profundus of the gastrocnemius.

Due to the diversity of pathologies and structures that myositis can affect, the diagnostic process for the disease may benefit from the extraction of multiple features or the analysis of the entire image. This was one of the many theories that were formalized after many hypotheses came before it. It is possible for there to be a "see-through" effect when there is edema, for instance, because there will be no perimysial echos. Even though the echo intensity has increased, it is still possible to see the bone beneath it. Among the possible manifestations of DM [10] are thickening of the fascia, inflammation of the subcutaneous tissue, and patchy involvement of the muscles. It may be easier to detect this type of structural change when examining the body's structure as a whole, as opposed to a single muscle.

Recent research on computer-aided diagnostics (CAD) has led to the development of muscle evaluations that are both more accurate and more reliable [11] than previously available ones. An operator may not always be able to consistently leverage, detect, and quantitatively analyze image biomarkers and features with computer algorithms. New ML techniques, such as deep learning (DL) may be helpful in computer-assisted myopathy diagnosis.

A condensed glossary of terms used by numerous authors in this field and subsequent works. The Artificial Intelligence (AI) subfield of computer science aims to develop intelligent software and hardware for computers. ML is a subfield of AI that focuses on developing computer algorithms capable of concluding data using statistical analysis [12]. Commonly, a statistical model is developed by constructing a model based on a collection of training data and then working backward. In the future, statistical techniques such as classification and regression may be employed to analyze newly collected data and draw conclusions based on this model [13, 14]. "DL," a subfield of ML, is distinguished from conventional ML by its reliance on neural networks, which are multilayered cascades of mathematical functions. This metric is what distinguishes DL from other ML strategies [15]. The network's ability to perform computations depends on millions of parameters, all of which are automatically uncovered by the system through training data, which contains labels already assigned to the appropriate category (es). In DL, convolutional neural networks, also known as deep convolutional neural networks (DCNNs) [16], are widely used for image analysis. This method convolves the input image with a small number of reusable filters. This reduces the astronomically large number of network parameters, making them more manageable.

An image feature set pertinent to the current problem at hand is manually selected and computed initially. The combined feature sets are then used to train and test a classifier using traditional ML methods such as Support Vector Machines (SVM) [18] or Random Forests (RF) [17–20]. There are 20 of them available in total. Using these more traditional approaches to the design of engineered features may result in a set of features that are either poorly selected or overly tailored to the training dataset. There is a possibility of either outcome occurring. As a result, the accuracy and generalizability of the model's predictions may remain unchanged. Conventional methods place an excessive amount of reliance on the knowledge and experience of the algorithmic designer when selecting these features. The features generated by DL techniques such as DCNNs [20] were neither designed nor selected by an engineer. These techniques instead generated these characteristics. DCNNs, are especially efficient at automatically learning image characteristics from data. As a result, DCNNs incorporate all processing pipeline stages into a single model, beginning with the most elementary and progressing to the most complex. This process involves the computation of features, the combination of features, and data classification at the very end. New technological and algorithmic advancements have led to significant improvements in the use of DCNN methods for general-purpose image classification. These advancements have led to these improvements. This progress is a result of the increased precision that has been achieved. For instance, the entire image can now be classified with the same precision as a human hand. Previously, this was impossible.

Electromyography can be used in conjunction with other diagnostic tools to determine whether or not a patient has muscular dystrophy. Electromyograms, also known as EMGs, are biomedical signals generated by muscle cells and characterized by an electrical potential. When these cells contract, an EMG is generated. EMG signals possess a low frequency (between 0 and 499 Hz) and a small amplitude compared to other types of signals. The EMG signal is predominantly comprised of frequencies between 50 and 150 hertz. EMGs permit the acquisition of a great deal of information regarding the nervous system and the anatomical and psychological properties of the muscles. Becker and Duchenne muscular dystrophy are, respectively, two of the disease's most prevalent forms. Early disease detection is essential for the overall effectiveness of the treatment plan. Using these data in future research could lead to the discovery of a treatment for the disorder as well as a better understanding of the disorder's nature. Normal EMGs and abnormal EMGs tend to have waveforms that are very similar to each other, making it hard to tell the difference between them. Even though it is necessary to do so in order to tell the difference between normal and abnormal EMG classes, the researcher is eager to process the signals. Even though it will take a lot of time and work, the researcher is eager to do it.

Due to its reliance on visual observation and the advice of experts, qualitative EMG analysis is prone to produce inaccurate results. The qualitative EMG analysis provides no data that can be used to compare or categorize various EMG disorders. It has been hypothesized that computer algorithms inspired by EMG could provide a solution to this issue [22]. According to research, various diagnostic approaches, including classification schemes and extracted feature sets, can be used to identify muscular disorders. Among these are the conventional bipolar EMG technique [21], linear and matrix electrode arrays [22], and various other classification techniques [23]. EMGs have been the subject of considerable research. The motor frequency coefficient of variation (MFCV), the motor unit size (MU), the frequency spectrum, and the entropy are examples of parameters. EMG features were extracted using a wavelet transform (WT) in [21], and the EMG data were then classified using a variety of neural networks. In [24], the Time domain method was used to analyze and classify EMG signals. The authors of [25] propose utilizing autoregressive and wavelet neural networks to extract a variety of EMG features Three distinct approaches were utilised to simplify the classification of electromyographic data: support vector machines (SVMs), decision trees (DTs), and random forest neural networks (RFNNs) (RBFNs) [26, 27, 28]. The paper [29] analyses an EMG signal using extracted parametric power spectral and WT features. A neuro-fuzzy classifier is utilized for this analysis. The classification accuracy of each of these suggested methods ranges from 72 percent to 86 percent. The only way to significantly improve classification accuracy is to conduct exhaustive research on the topic. The accuracy of this study's findings should be greater than ninety percent.

6.3 EA FOR DIAGNOSING MUSCULAR DYSTROPHY

Myopathy is a disease that exclusively impacts the muscles and the tissues that are responsible for their support. It does not move to other body areas and does not spread. Myopathic symptoms include muscle weakness, stiffness, cramping, and spasms, as well as muscle dysfunction, cramping, and spasms.

Other myopathic symptoms include muscle dysfunction, cramping, and spasms. However, based solely on a person's symptoms, it can be extremely challenging to diagnose myopathy [13]. The study of the electrical signals that are sent by the nervous system is called electromyography. Both invasive and non-invasive methods can be used to measure electromyographic signals. On the other hand, if you only look at the EMG signals, you won't be able to figure out what is wrong because normal and abnormal signals may have the same appearance. As a result, extracting helpful and informative features from EMG signals is necessary to develop diagnostic tools and systems. This is essential for the development of these tools and systems. In order to complete the project, you must do this.

Extracting and selecting variable feature vectors and constructing classifier models are necessary to distinguish between normal and abnormal EMG signals. Standard EMG signals can only be distinguished from abnormal signals in this manner. There are three ways to extract biological signals' frequency, time, and frequency-time features, which can be done in both standard and abnormal circumstances. In order to accurately analyze an EMG signal, factors other than frequency and time must be considered. Thus, characteristics that consider both time and frequency are referred to as time-frequency characteristics.

Genetic algorithm (GA)-based ML and DL framework for dystrophy classification is shown in Figure 6.2. Time-frequency transformation methods [25] can be used to extract various time-frequency features from EMG signals. These characteristics can be used to monitor muscle activity. The

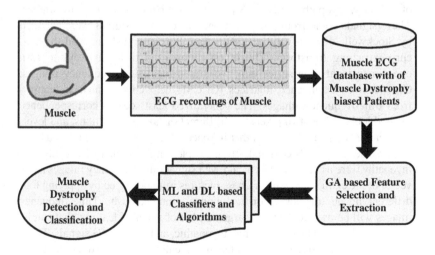

FIGURE 6.2 GA-based ML and DL framework for dystrophy classification.

time-frequency properties of signals must be known in order to analyze them. The reason for this is that as time passes, the signal's frequency changes. When creating automated diagnostic tools, we must distinguish between normal and abnormal EMG signals. A classifier model had to be built, and the appropriate feature subsets had to be acquired in order to distinguish between normal and abnormal signals. Researchers have used a variety of methods, such as Stockwell, Synchro-Extracting, Wigner–Ville, and Short-Time Fourier Transform, to extract the time-frequency characteristics of EMG signals from healthy people and people with ALS. EMG signals from both healthy and ALS patients were analyzed using these methods.

The GA [31] was utilized to narrow the list down to the 15 most critical characteristics. Using three distinct classification models for healthy individuals and ALS patients, we examined the statistical significance of a few of these characteristics (one for each). In addition to the GA-selected features, we also used artificial neural networks to build classifiers for the extracted transformed time-frequency features. EMG classifiers can be constructed utilizing features selected by a genetic algorithm based on the study's results.

In order to record normal, myopathy, and ALS, concentric needle electrodes are typically utilized for recording (EM) signals. These EMG signals can be located in an open database of EMGs. Replaying these signals can be done with the help of an EMG machine. The adopted EMG signals were recorded at a sampling rate, and their source was the electromyographic activity that was produced by the brachial biceps muscles. These values will be obtained following a database search containing EMG signal information for both standard and abnormal readings. Time-frequency EMG images of ordinary, myopathic, and ALS patients have been made in many studies. These signals come from more than one place. These signals were changed by Stockwell, Wigner–Ville, the short-time Fourier transform, and the synchro-extracting transform. The investigation used different ways to turn one-dimensional time series into two-dimensional time-frequency images. Gray-level co-occurrence matrices (GLCM) help choose surface features from time-frequency images. The generalized least squares correspondence method looks at how two pixels with different distances, directions, and levels of variation are related to each other in space [30]. This method gets data from a grayscale image. EMG signals from people with normal EMG, ALS, and myopathy were used to make typical EMG signal time-frequency images and a vector of the extracted features. These pictures were produced by applying four different image transformation techniques simultaneously. Time-frequency images will be used to extract a large number of valuable features that can be used to analyze and classify normal, myopathic, and ALS EMG signals. These features will be extracted using a technique called feature extraction. These factors are going to be considered in the analysis. The genetic algorithm will select a subset of those features that contain useful information from all others to solve the problem. This subset will be chosen at random. In the future, machine learning and deep learning classifiers will be created using the most informative features and extracted time-frequency features to differentiate between normal, myopathic, and ALS cases. These classifiers will use the most informative features and extracted time-frequency features.

The authors [32] came up with a method to determine whether or not a person has DMD by combining a Raman hyperspectroscopic analysis of blood serum with a more sophisticated statistical analysis. This created a test that could determine whether or not a person has DMD. The combination was undertaken to achieve more precise results. A pattern-finding algorithm known as PLSA was applied to the serum spectral data of mice with DMD as well as mice without the condition. This was done in order to make the patterns more visible (MDX). According to the findings of the cross-validation, diseased spectra can be identified with a sensitivity of 95.2 percent and a specificity of 94.6 percent.

The researchers [33] used genetic programming (GP) to generate a list of potential adjustments that could be made to the space occupied by grayscale images. The term "transform-based evolvable feature" (TEF) was coined to describe their concept within the business world context. A moment value is assigned to an image that has been GP-transformed to be utilized in a classification activity. Because of this, it will be possible to use the image more precisely. The TEF makes it possible to search and improve the entire image space, which was impossible with many of the methods that came before it. These other methodologies were developed in the past, before the TEF. TEFs are constructed through the utilization of Cartesian GP, and then they are applied to a medical image classification task, specifically the detection of inclusions in cell images that indicate MD. This particular task focuses on the detection of inclusions in cell images. The task at hand is a medical image classification task.

Researchers from [34] looked into the possibility of using next-generation sequencing, also known as NGS, to improve diagnostics for patients whose samples lacked specific molecular characterization. Hereditary myopathies were the primary focus of this particular research. Individualized NGS may benefit patients with early-onset myopathies and MDs, such as collagen VI-related myopathy and congenital myasthenia syndromes. NGS can complete sequencing tasks in days or weeks, while traditional methods take months or years. Because of this, the care that these patients receive will be of a higher standard. In spite of this, a significant portion of the population that we were studying did not have a molecular diagnosis. This could be because NGS has inherent limitations in its ability to detect specific types of mutations, or it could be because additional genes responsible for neuromuscular disorders have not yet been identified. Both of these possibilities are possible. The results could go either way in both of these scenarios.

A discussion on GP as a diagnostic tool for muscular disorders can be found in [35]. The utilization of biosignals was a topic of discussion. The electrical activity that is produced by particular cells, such as those found in the heart, the nervous system, and the muscles, is an important aspect of a biosignal. Because it provides information on the flow of data through the nervous system and the activation of various muscles as a result, the study of muscle signals is essential for understanding the function of the system. This is the case for the reasons that were discussed earlier. The abnormal electrical activity that is characteristic of neuropathies, MDs, and sclerosis can all be detected by this signal. Other potential uses for this signal include control systems for robotics, prosthetics, and even telemedicine, to name just a few of the many possible applications. The signals produced by electromyography are characterized as chaotic, non-stationary, and non-linear. As a direct consequence, the most recent generation of EMG pattern recognition systems incorporates modules for preprocessing, segmentation, feature extraction, dimensionality reduction, classification, and data control [36]. It is essential to locate a pattern that accounts for the entirety of the signal rather than focusing on the signal's data points.

The early onset of these conditions, which are collectively referred to as CMDs, as well as the presence of histological characteristics that are suggestive of a dystrophic process is what set them apart from other conditions. Due to the clinical and genetic heterogeneity that exists within the group that is comprised of congenital MDs, it is becoming increasingly difficult to obtain an accurate genetic diagnosis, even in this day and age when NGS technology is available. In order to arrive at an accurate molecular diagnosis, the researchers investigated the diagnostic characteristics, differential diagnostic considerations, and available diagnostic tools for each of the different CMD subtypes. This allowed the researchers to arrive at an accurate molecular diagnosis. After that, they offered an instruction manual on how to use these resources methodically.

6.4 CONCLUSION

Researchers had to rely heavily on clinical data and observations, which represented a significant barrier, in order to classify and predict MD using EMG signals and imaging. This presented a challenge for the researchers. The researchers were faced with a difficult task as a result of this. Historically, EMG datasets were categorized using machine learning or deep learning algorithms, depending on whether they were being represented in the frequency or time domain. The conventional method does not result in accurate classification while also being efficient concerning both time and labor. It has been found that EA algorithms are a robust optimization method that can be utilized in selecting critical diagnostic features for MD. This discovery came about as a result of the fact that EA algorithms are practical. In this chapter, various electromyographic signals and imaging of the muscles were utilized to investigate the various genetic approaches to diagnosing MD.

REFERENCES

1. Narasimhaiah, D., Uppin, M. S., & Ranganath, P. (2022). Genetics and muscle pathology in the diagnosis of muscular dystrophies: An update. *Indian Journal of Pathology and Microbiology*, *65*(5), 259.
2. Mohamadian, M., Rastegar, M., Pasamanesh, N., Ghadiri, A., Ghandil, P., & Naseri, M. (2021). Clinical and molecular spectrum of muscular dystrophies (MDs) with intellectual disability (ID): A comprehensive overview. *Journal of Molecular Neuroscience*, *72*, 9–23.
3. Bădilă, E., Lungu, I. I., Grumezescu, A. M., & Scafa Udrişte, A. (2021). Diagnosis of cardiac abnormalities in muscular dystrophies. *Medicina*, *57*(5), 488.
4. Roy, B., & Griggs, R. (2021). Advances in treatments in muscular dystrophies and motor neuron disorders. *Neurologic Clinics*, *39*(1), 87–112.
5. Fortunato, F., Rossi, R., Falzarano, M. S., & Ferlini, A. (2021). Innovative therapeutic approaches for duchenne muscular dystrophy. *Journal of Clinical Medicine*, *10*(4), 820.
6. Aivazoglou, L. U., Guimarães, J. B., Link, T. M., Costa, M. A. F., Cardoso, F. N., de Mattos Lombardi Badia, B. ... & da Rocha Corrêa Fernandes, A. (2021). MR imaging of inherited myopathies: a review and proposal of imaging algorithms. *European Radiology*, *31*(11), 8498–8512.
7. Pino, M. G., Rich, K. A., & Kolb, S. J. (2021). Update on biomarkers in spinal muscular atrophy. *Biomarker Insights*, *16*, 11772719211035643.
8. Alfano, L. N., Focht Garand, K. L., Malandraki, G. A., Salam, S., Machado, P. M., & Dimachkie, M. M. (2022). Measuring change in inclusion body myositis: Clinical assessments versus imaging. *Clinical and Experimental Rheumatology*, *40*(2), 404–413.
9. Stock, M. S., & Thompson, B. J. (2021). Echo intensity as an indicator of skeletal muscle quality: Applications, methodology, and future directions. *European Journal of Applied Physiology*, *121*(2), 369–380.
10. Wijntjes, J., & van Alfen, N. (2021). Muscle ultrasound: Present state and future opportunities. *Muscle and Nerve*, *63*(4), 455–466.
11. Bazaga, A., Roldán, M., Badosa, C., Jiménez-Mallebrera, C., & Porta, J. M. (2019). A convolutional neural network for the automatic diagnosis of collagen VI-related muscular dystrophies. *Applied Soft Computing*, *85*, 105772.

12. Bowden, S. A., Connolly, A. M., Kinnett, K., & Zeitler, P. S. (2019). Management of adrenal insufficiency risk after long-term systemic glucocorticoid therapy in Duchenne muscular dystrophy: Clinical practice recommendations. *Journal of Neuromuscular Diseases, 6*(1), 31–41.
13. Cai, J., Xing, F., Batra, A., Liu, F., Walter, G. A., Vandenborne, K. ... & Yang, L. (2019). Texture analysis for muscular dystrophy classification in MRI with improved class activation mapping. *Pattern Recognition, 86*, 368–375.
14. Crisafulli, S., Sultana, J., Fontana, A., Salvo, F., Messina, S., & Trifirò, G. (2020). Global epidemiology of Duchenne muscular dystrophy: An updated systematic review and meta-analysis. *Orphanet Journal of Rare Diseases, 15*(1), 1–20.
15. Ramli, A. A., Zhang, H., Hou, J., Liu, R., Liu, X., Nicorici, A. ... & Henricson, E. (2021). Gait characterization in Duchenne muscular dystrophy (DMD) using a single-sensor accelerometer: Classical machine learning and deep learning approaches. *arXiv preprint arXiv:2105.06295.*
16. Li, Y., Yang, Z., Wang, Y., Cao, X., & Xu, X. (2019). A neural network approach to analyze cross-sections of muscle fibers in pathological images. *Computers in Biology and Medicine, 104*, 97–104.
17. Shin, Y., Yang, J., Lee, Y. H., & Kim, S. (2021). Artificial intelligence in musculo-skeletal ultrasound imaging. *Ultrasonography, 40*(1), 30.
18. Kehri, V., & Awale, R. N. (2018). Emg signal analysis for diagnosis of muscular dystrophy using wavelet transform, svm and ann. *Biomedical and Pharmacology Journal, 11*(3), 1583–1591.
19. Alfano, L. N., & Mozaffar, T. (2021). Random forest: Random results or meaningful insights for patients with facioscapulohumeral muscular dystrophy? *Brain, 144*(11), 3288–3290.
20. Konda, A., Crump, K., Podlisny, D., Meyer, C. H., Blemker, S. S., Hart, J. & Feng, X. (2018). Fully automatic segmentation of all lower body muscles from high resolution MRI using a two-step DCNN model. Proceedings of 26th Annual Meeting ISMRM, Paris, France, 1398.
21. Nizamis, K., Rijken, N. H., Van Middelaar, R., Neto, J., Koopman, B. F., & Sartori, M. (2020). Characterization of forearm muscle activation in duchenne muscular dystrophy via high-density electromyography: A case study on the implications for myoelectric control. *Frontiers in Neurology, 11*, 231.
22. Rinaldi, M., Petrarca, M., Romano, A., Vasco, G., D'Anna, C., Schmid, M. ... & Conforto, S. (2019, July). EMG-based indicators of muscular co-activation during gait in children with duchenne muscular dystrophy. In *2019 41st Annual international conference of the IEEE Engineering in Medicine and Biology Society (EMBC)* (pp. 3845–3848). Berlin, Germany: IEEE.
23. Heller, S. A., Shih, R., Kalra, R., & Kang, P. B. (2020). Emery-Dreifuss muscular dystrophy. *Muscle & Nerve, 61*(4), 436–448.
24. Askarinejad, S. E., Nazari, M. A., & Borachalou, S. R. (2018, March). Experimental detection of muscle atrophy initiation Using sEMG signals. In *2018 IEEE 4th Middle East Conference on Biomedical Engineering (MECBME)* (pp. 34–38). Tunisia, Africa: IEEE.
25. Liu, Z., Wang, X. A., Su, M., & Le, L. (2019). Research on rehabilitation training bed with action prediction based on NARX neural network. *International Journal of Imaging Systems and Technology, 29*(4), 539–546.

26. Toledo-Pérez, D. C., Rodríguez-Reséndiz, J., Gómez-Loenzo, R. A., & Jauregui-Correa, J. C. (2019). Support vector machine-based EMG signal classification techniques: a review. *Applied Sciences*, *9*(20), 4402.
27. Altan, E., Pehlivan, K., & Kaplanoğlu, E. (2019, April). Comparison of EMG based finger motion classification algorithms. In *2019 27th Signal Processing and Communications Applications Conference (SIU)* (pp. 1–4). Sivas, Turkey: IEEE.
28. Lv, Z., Xiao, F., Wu, Z., Liu, Z., & Wang, Y. (2021). Hand gestures recognition from surface electromyogram signal based on self-organizing mapping and radial basis function network. *Biomedical Signal Processing and Control*, *68*, 102629.
29. Shanmuganathan, V., Yesudhas, H. R., Khan, M. S., Khari, M., & Gandomi, A. H. (2020). R-CNN and wavelet feature extraction for hand gesture recognition with EMG signals. *Neural Computing and Applications*, *32*(21), 16723–16736.
30. Verhaart, I. E., & Aartsma-Rus, A. (2019). Therapeutic developments for Duchenne muscular dystrophy. *Nature Reviews Neurology*, *15*(7), 373–386.
31. Khan, M. U., Aziz, S., Bilal, M., & Aamir, M. B. (2019, August). Classification of EMG signals for assessment of neuromuscular disorder using empirical mode decomposition and logistic regression. In *2019 International Conference on Applied and Engineering Mathematics (ICAEM)* (pp. 237–243). Taxila, Pakistan: IEEE.
32. Ambikapathy, B., Kirshnamurthy, K., & Venkatesan, R. (2021). Assessment of electromyograms using genetic algorithm and artificial neural networks. *Evol. Intel. 14*, 261–271. https://doi.org/10.1007/s12065-018-0174-0
33. Ralbovsky, N. M., Dey, P., Galfano, A & Dey, B. K. (2020). Diagnosis of a model of Duchenne muscular dystrophy in blood serum of *mdx* mice using raman hyper-spectroscopy. *Sci Rep*, *10*, 11734. https://doi.org/10.1038/s41598-020-68598-8
34. Kowaliw, W., Banzhaf, N., Kharma, S., & Harding. (2009). Evolving novel image features using genetic programming-based image transforms. *2009 IEEE Congress on Evolutionary Computation*, pp. 2502–2507. https://doi.org/10.1109/CEC.2009.4983255.
35. Chae, J. H., Vasta, V., Cho, A., Lim, B. C., Zhang, Q., Eun, S. H. … & Hahn, S. H. (2015). Utility of next generation sequencing in genetic diagnosis of early onset neuromuscular disorders. *Journal of Medical Genetics*, *52*(3), 208–216.
36. Aviles, M., Sanchez-Reyes, L.-M., & Toledo-Pérez, D. C., (2021). A novel methodology to classify myoelectric signals using genetic algorithms and support vector machines. *2021 XVII International Engineering Congress (CONIIN)*, pp. 1–8. https://doi.org/10.1109/CONIIN54356.2021.9634836.

Tumor Prediction and Classification

7

7.1 INTRODUCTION

In the medical community, the term "cancer" describes a wide range of diseases characterized by the uncontrollable division of abnormal cells and the ability to invade and destroy normal body tissue. In addition, cancer can develop in several different locations throughout the body. Metastasis, or the spreading of cancer to other body parts, is possible. According to the World Health Organization (WHO), cancer is the second leading cause of death worldwide. Many types of cancer now have higher survival rates due to advancements in cancer screening, treatment, and prevention. Cells in the human body grow in a predictable pattern under the guidance of a complex regulatory system. A lump or tumor is formed when a group of cells suddenly and uncontrollably begin to multiply. This process characterizes cancer. There is a risk that a tumor will spread to other parts of the body if it continues to grow out of control [1]. Every cell in the body is associated with a fixed life span, after which it dies naturally, which is termed as apoptosis. Apoptosis is a process by which a cell receives an instruction to die which is thereafter replaced by a newer cell that functions comparatively in a better manner [2]. The cancerous cells are ones that lack the component needed to stop the whole process of cell division and death, making them carry on with their division activity for a prolonged lifetime. These tissues and cells accumulated in the body take the form of body outgrowths referred to as

DOI: 10.1201/9781003254874-7

lumps that absorb the nutrients and oxygen, which are meant for other cells. Cancerous cells have the potential to form tumors, are capable of destroying the immune system, and may cause changes that stop the body from functioning normally.

There are numerous methods of treating cancer, such as chemotherapy, immunotherapy, radiation therapy, stem-cell transmission, and surgery. Apart from these, there are other disease-specific treatments that may be adopted with other therapies in combination, in order to maximize the effect on the cancer cells. The doctors usually prescribe the treatment depending on the stage of the cancer along with the person's overall health.

The cancer cells begin at one part, spread out to other parts with time, affecting the whole body gradually, via lymph nodes i.e. the cluster of immunization cells located all over the body. The spreading of cancer cells to other parts is medically termed as metastasis, which can be prevented by early treatment, making the need of early cancer detection all the more vital as a contribution to improve the cancer survival rate all over the world. This chapter discusses different facets of cancer detection, followed by various evolutionary intelligence algorithms that support or claim to facilitate an early detection of cancer of various kinds.

7.2 TUMOR TYPES

A tumor is a mass or group of abnormal cells that can form inside the body. Even if you have a tumor, that does not mean you have cancer. Many kinds of tumors are harmless (not cancerous). Tumors can start growing in any part of the body. They might affect the bones, skin, tissues, glands, and organs. Neoplasm is another word for a tumor. A tumor can be put into one of three different groups [3].

7.2.1 Carcinogenic

Tumors that are malignant or cancerous can spread to nearby tissue, glands, and other body parts. This is called metastasis. The new tumors have grown from metastases. Malignant tumors may come back in a patient after treatment. These tumors could endanger a person's life.

Malignant tumors have cells that multiply out of control and can spread locally or to other body parts. Malignant tumors are cancerous (i.e., they invade other sites). They could get to faraway places by going through the bloodstream or the lymphatic system. Metastasis is the name for this process. Any part of the body can show signs of metastasis, but the liver, lungs, brain, and bones are the most common places for this to happen [4].

Treatment is needed to stop the malignant tumor from spreading. Surgery, followed by either chemotherapy or radiation therapy, is the best way to treat cancers in their early stages. If cancer has spread to other body parts, the best treatment will probably be one that works throughout the body, like chemotherapy or immunotherapy.

7.2.2 Noncancerous

Tumors that are not cancerous and rarely risk the patient's life are called benign. They are localized, meaning they usually do not affect nearby tissue and do not spread to other body parts. Many tumors are harmless and do not need to be treated. On the other hand, some benign tumors can put pressure on other parts of the body and need medical help.

Benign tumors don't spread to other parts of the body and stay in the place where they started. When cancerous tumors grow, they spread to other parts of the body. They don't move to other buildings or parts of the body outside of the affected area. Tumors that are not harmful usually grow slowly and have precise edges.

Most of the time, benign tumors do not cause any harm. However, they can get big and press on nearby structures, which can be painful and cause other health problems. For example, a sizeable benign lung tumor can pressure the trachea (also called the windpipe), making breathing hard. Because of this, it needs to be taken out by surgery right away. Benign tumors rarely come back after surgery is done to remove them. Common tumors that are not cancerous include fibroids in the uterus and lipomas on the skin.

Some types of benign tumors can change into cancerous tumors. These are carefully watched and may need to be taken out surgically in the future. For example, colon polyps, just another name for a group of abnormal cells, can turn into cancer and are usually removed through surgery [5].

7.2.3 Precancerous

Premalignant cells are cells that have not yet become cancer. They are abnormal cells that could turn into cancer cells, but they do not invade or spread on their own. If these non-cancerous tumors are not treated, they could become cancerous ones. Sometimes it can be hard to understand what precancerous cells are and whether or not they will turn into cancer. This is because there is no clear answer to the question. Most of the time, cells do not go from being normal on day one to becoming precancerous on day two and then cancerous on day three. Usually, this process takes more than three days.

It's important to say again that precancerous cells are not the same as cancer cells. This means that if they are left alone and not treated, they won't spread to other body parts. They are just abnormal cells that, if given enough time, might change in ways that turn them into cancer cells. If the cells are killed before turning into cancer cells, the disease should be utterly curable, at least in theory. Still, not all cells that could become cancerous need to be killed immediately. Another thing that confuses people is that many tumors have cancer cells and cells that are getting ready to become cancer in the same tissue. For example, some people with breast cancer have cancer cells in a tumor, but there may also be precancerous cells in other parts of the breasts or even in the tumor itself [6]. There are also precancerous cells in the tumor.

7.3 CARCINOMA CLASSIFICATION

For example, mucous membranes and gastrointestinal tracts are epithelial tissues, which is where this occurs. According to the National Cancer Institute, carcinomas make up between 80 percent and 90 percent of all cancer cases.

In the body's bone marrow, cancer called leukemia develops, which is responsible for producing blood cells. The thymus, tonsils, and spleen are all part of the lymphatic system, where lymphomas start. The immune system and hormones are intertwined with the operation of this system. There are two main categories of mixed cancer: those that affect both normal and cancerous cells and those that simultaneously affect both normal and cancerous cells. Plasma cells in the body's circulatory system begin to mutate, and myeloma is the most common form of the disease.

Bones, muscle, fat, and cartilage can all be affected by sarcomas, which originate in connective tissue and spread to other tissues and organs. Younger people are more likely to develop sarcomas than older people. A doctor can distinguish one cancer from another based on the appearance of cancer cells. Doctors benefit from a thorough understanding of the various types of cancer when developing a treatment plan [6].

7.3.1 Lung Carcinoma

Lung carcinoma is the world's third most common type of cancer, accounting for approximately 3 percent of all cases. Is one of the most invasive forms of cancer, spreading to other parts of the body quickly and frequently. In both sexes, it is the most deadly form of cancer. Around 2.21 million people will be affected by lung carcinoma in the world by 2020, according to the WHO. It is the leading cause of cancer-related death yearly, killing more people than any other single factor. Long-term smoking is linked to an increased risk of developing lung cancer and a higher mortality rate from the disease. There may be a link between a person's ability to repair DNA damage and their risk of developing lung cancer from smoking, as well as gene-enzyme interactions that influence the activation or detoxification of carcinogens and the formation of DNA adducts. In addition, the level of DNA adduct formation may influence an individual's susceptibility. A diverse disease, lung carcinoma can originate in many places within the bronchial tree. Because of this, the symptoms and signs of lung carcinoma can vary greatly depending on the disease's anatomical location. If a patient is diagnosed with stage III or stage IV of the disease, he or she is 70 percent likely to survive.

There are approximately 25–30 percent of all lung cancers that are SQCLCs. These cancers usually begin in the main bronchi and progress to the carina. Forty percent of all lung cancers originate in the periphery of the bronchi, where they are known as adenocarcinomas (AdenoCA). Lobar atelectasis and pneumonitis are two hallmarks of the progression of AdenoCAs. Interalveolar connections between the alveoli and the rest of the body are the primary means by which adenocarcinoma (AIS) and minimally invasive adenocarcinoma (MIA) spread from alveoli to the rest of the body. When complete resection is performed, patients with an AIS or MIA designation are considered to be in excellent health.

A visual representation of the non-small lung cancer staging system (NSCLC). To be clear, squamous and adenocarcinoma originate in the main and lobar bronchi, respectively, while adenocarcinomas originate in the peripheral lung tissue and the epithelial cells of the segmental bronchi, respectively. For non-small cell lung cancers in stage IV, the rate of distant metastasis to non-thoracic organs is shown. There is a correlation between the percentage of patients with squamous cell lung cancers and those with adenocarcinomas who have metastasized.

In contrast to other types of cancer, small cell lung cancers (SCLC) originate from the hormone-producing cells of the lung and typically present as central mediastinal tumors. All lung cancers, including SCLC, account for about 15 percent of all cases. Approximately 10–15 percent of all lung cancer cases are SCLC, a subtype of the disease. In contrast to small cell anaplastic carcinomas, large cell anaplastic carcinomas, or NSCLC NOS, are more proximal in location and have a tendency to invade the mediastinum and its structures early on. This is because NSCLC-NOS, are less likely to invade the bronchial tree.

NSCLC-NOS accounts for about 10 percent of all NSCLC cases, and like SCLC, it spreads rapidly and is fatally so. In the superior sulcus, the pancoast cancer begins and then spreads throughout the body by invading adjacent structures. Tumors of the lungs can become multifocal, spread to the opposite lung, or even spread to the opposite side, depending on where they first appeared (referred to as "reverse metastasis" in this context) (M1). When lymph nodes in the mediastinum are compressed, it can result in esophageal compression and difficulty swallowing, venous compression and congestion due to collateral circulation, or tracheal compression. Mediastinum compression is always linked to advanced lymph node involvement. Advanced lymph node involvement is almost always present when the mediastinum is compressed. The presence of metastatic disease in distant organs such as the liver, brain, or bone can be detected long before a primary lung lesion is discovered [7].

7.3.2 Blood Carcinoma

Many types of blood cancers can affect the blood cells and the "spongy" bone marrow that produces blood cells. This type of cancer alters the function and behavior of blood cells. Human blood cells can be classified into three types [9]. White blood cells play a critical role in the body's ability to fight off infection. Red blood cells' job is to transport oxygen throughout your body and carry carbon dioxide to your lungs, where it is exhaled. Platelets aid in the formation of blood clots when you are injured. One of the most common forms of blood cancer is acute myeloid leukemia [8].

7.3.2.1 Multiple myeloma

In the case of this type of cancer, your bone marrow, and lymphatic system produce blood cells that don't function properly. Depending on the type of white blood cell, they all have a different effect on them.

7.3.2.2 Leukemia

In patients with leukemia, the production of white blood cells is abnormally high, making them unable to fight off infections. The type of white blood cell affected and the rate at which the disease progresses (acute vs. chronic) distinguish the four different subtypes of leukemia [9, 10].

- Acute lymphocytic leukemia (ALL): This process is initiated by lymphocytes, white blood cells in the bone marrow. A high number of lymphocytes produced by people with ALL can inhibit the production of healthy white blood cells. If you don't get treatment, ALL can get worse very fast.
- It's the most common type of childhood cancer. Adults can also get ALL, despite the fact that it is more common in children between the ages of 3 and 5.
- Acute leukemia of the blood (AML): These cells, which generally mature into white blood cells, red blood cells, or platelets, are at the root of this condition. In patients with AML, the incidence of all three types of healthy blood cells is decreased. Rapidly progresses this form of leukemia. There is a higher incidence of AML in people 65 and older. Men are more susceptible than women to this condition.
- Chronic lymphocytic leukemia (CLL): Adults are more likely to be diagnosed with this type of leukemia than any other subtype. It begins in the bone marrow lymphocytes, like ALL, but it progresses more slowly. People with CLL typically don't show any symptoms until years after they've already been diagnosed.
- Cancerous lymphoma (CLL) is most commonly found in people in their 70s and older. In addition to having a history of blood cancer in your family, exposure to chemicals such as weedkillers and insecticides can raise your risk of developing the disease.

For those with a chronic form of myeloid leukemia (CML). Similar to AML, this type of blood cancer begins in myeloid cells. However, the growth of abnormal cells is slow.

- Men are more likely than women to be diagnosed with CML. Adults are the most common victims, but children of any age can fall prey to its snares.

7.3.2.3 Lymphoma

This is a cancerous disease that affects the lymphatic system. Within this network of blood vessels are your lymph nodes, spleen, and thymus gland. White blood cells are stored in the vessels and circulated throughout the body to help fight infection. Lymphoma begins in white blood cells, which are also known as lymphocytes. In terms of lymphoma, there are two main categories, which are as follows: B lymphocytes, or simply B cells, are the immune cells that give rise to Hodgkin's lymphoma [11]. They are proteins made by these cells that protect the body from pathogens. Antibodies are a type of antibody. Reed-Sternberg cells are large lymphocytes found in the lymph nodes of people with Hodgkin's lymphoma. Non-lymphoma A type of immune cell known as a T cell can also be a source of Hodgkin's disease. Non-lymphoma Hodgkin's is far more common than Hodgkin's lymphoma in the general population.

Each type has a few distinct subtypes that can be further broken down. Depending on how cancer behaves and where it first manifests itself in the body, a subtype can be assigned to it. Individuals with weakened immune systems are more likely to develop lymphoma. In addition, having Epstein-Barr virus, HIV, or Helicobacter pylori (H. pylori) infection increases your risk. A lymphoma diagnosis is most commonly made in people ages 15 to 35 (as well as those ages 50 and older).

7.3.2.4 Myeloma

The bone marrow's plasma cells are affected by this disease. In the body, antibodies are produced by plasma cells, a type of white blood cell. The bone marrow is filled with cancerous cells from the myeloma virus. Bones can be damaged and they can obstruct healthy blood cells. Additionally, the antibodies produced by these cells are ineffective in protecting the body against infection. It's called multiple myeloma because it can spread to multiple parts of your bone marrow, and that's where the name comes from. Men of African-American descent are more likely than men of any other race or ethnicity to develop the disease, especially after the age of 50.

7.3.3 Colon Carcinoma

The cancer can spread to the colon when the large intestine becomes infected. The colon is located at the very end of the digestive system, at the furthest point from the mouth. This cancer is more common in the elderly, despite the fact that it can affect anyone at any time. Polyps, which are noncancerous clusters of cells found in the lining of the colon, are commonly thought to be the origin of the problem. If these polyps are left untreated, they may develop into colon cancer. Potentially small polyps may have few or no symptoms. Colon cancer risk can be reduced through routine screening tests, which can detect polyps and remove them before they turn into cancerous tumors. Colon cancer can be controlled and kept under control with a variety of treatments, including surgery, radiation therapy, chemotherapy, targeted therapy, and immunotherapy [12].

Colorectal cancer, a term that combines the terms "colon cancer" and "rectal cancer," refers to cancer that begins in the rectum. Colon cancer is sometimes referred to as cancer of the colon. Most cases of colon cancer are caused by genetic mutations, which alter the DNA of healthy cells in the intestine. A cell's DNA contains a set of instructions that tell it what it should be doing at any given time. To keep your body functioning normally, healthy cells must grow and divide in a regulated manner. The cell will continue to divide even though it doesn't need any new cells if its DNA is damaged and it develops cancer. As the cells begin to gather, a tumor will form. The cancer cells can spread to other tissues over time, causing their destruction. Additional to that, cancerous cells can spread throughout the body and form deposits in new locations, which is known as metastasis (metastasis) [13].

7.3.4 Bone Cancer

Bone cancer occurs when abnormal bone cells proliferate unchecked. Normal bone tissue is destroyed. One of your bones may be the source of the disease, or it may spread there from elsewhere in your body (called metastasis). Bone cancer is a rare occurrence. There are a large number of benign tumors that form in bones, which means that they are not cancerous and do not spread to other areas of the body. But they still have the potential to cause bone fragility, which can lead to fractures and more serious health problems. In terms of benign bone tumors, there are a few main categories of tumors that fall into this category:

Osteochondroma is the most common name for this condition. The majority of those affected are under the age of twenty. The leg is the most common site of a giant cell tumor. These can also become cancerous in extremely rare circumstances. People in their twenties and thirties are most likely to develop osteoid osteoma, which affects the long bones. Osteoblastoma is a relatively rare tumor that typically affects young adults and grows in the spine and long bones. It is most commonly found in the spine.

- Hands and feet are the most common sites where enchondroma appears. In many cases, there are no symptoms at all. The most common type of tumor in the hand, it is called carcinoma.

7.3.4.1 Bone cancer that is the result of an undiagnosed malignancy

Primary bone cancer, or bone sarcoma, is a malignant tumor that originates in the bones. There is no known cause for it, but genetics may play a role. Some of the most common types of bone cancers are listed here:

- Knee and upper arm osteosarcoma are the most commonly affected areas. Adults with Paget's bone disease may also be at risk for Ewing's sarcoma, which is more common in people between the ages of 5 and 20 years, but it can also affect younger people. Sheath hernias are most commonly found in the rib cage; the upper arm; and the lower leg or foot. It's also possible that the soft tissue that covers your bones is the source of the infection.

Chondrosarcoma is diagnosed most frequently in patients between the ages of 40 and 70 years. The hip and pelvis, as well as the leg, arm, and shoulder, are frequently affected by this cancer, which originates in the cartilage cells of your body.

- Despite the fact that it begins in the bones, multiple myeloma is not considered a primary bone cancer. The soft tissue found within bones, the marrow, is affected by this form of cancer.
- For the vast majority of people, cancer that spreads to their bones began in another part of their body first. An example of secondary bone cancer is when lung cancer has spread to the bones, resulting in secondary bone cancers. "Metastatic cancer" refers to cancer that has spread from one part of the body to another.

7.3.5 Liver Carcinoma

- Those with chronic liver disease or cirrhosis are more likely to develop a malignant tumor known as liver cancer, which begins in the liver. An aggressive tumor, liver cancer is. It is the seventh most common form of cancer in women and the fifth most common form of cancer in men, hepatocellular carcinoma (HCC). Even more so, it is the third most common reason people die from cancer internationally. A yearly average of more than two times as many men as women are told they have liver cancer. We see this disparity all over the world. HCC is rare in South-Central Asia, Western Asia, Northern Europe, and Eastern Europe. East and Southeast

Asia, as well as the Middle and Western Africa, have the highest risk of liver cancer.

- In addition to being the body's largest organ, the liver is also its largest gland. HCC, which is the most common form of liver cancer, is the sixth most common form of the disease and the third most common cause of cancer-related death in the United States overall. Liver cancer is the second leading cause of cancer-related death in men worldwide.

- Liver cancer can be detected using various image-based diagnostic techniques such as ultrasonography, angiography, MRI, and computed tomography (CT) scans. CT scan images are preferred for early diagnosis because of their high resolution and the fact that they are non-invasive to the human body. Using CT scans, liver lesions can be distinguished by comparing their pixel intensities. The manual segmentation of CT scans is a time-consuming and tedious process that requires a lot of effort. As a result, automatic segmentation is the best option because it will reduce diagnostic time and benefit patients [14].

7.3.6 Bladder Carcinoma

Bladder cancer is one of the most common types of cancer, and it is caused by the cells in the bladder. The bladder, a hollow muscular organ in the lower abdomen, serves as a storage facility for urine. Urothelial cells, which line the bladder's interior, are where bladder cancer typically begins. Your kidneys and the tubes (ureters) that connect your kidneys to your bladder contain urothelial cells. On rare occasions, urothelial cancer may develop in the kidneys or ureters, but this rarely happens. In most bladder cancer cases, the disease is discovered at an early stage, when it is most treatable. In spite of this, even bladder cancers that were diagnosed and successfully treated at an early stage can recur. In order to check for a recurrence of the disease, patients who have had treatment for bladder cancer typically require follow-up tests for several years after their treatment has ended [15].

7.3.6.1 Diagnosis of cancer of the bladder

Diagnostic procedures for bladder cancer include the following tests and procedures:

The scope examination of the bladder's internal organs (cystoscopy). During a cystoscopy procedure, your doctor inserts a cystoscope into your urethra and uses it to view the inside of your bladder. Because of the lens on the cystoscope, your doctor can examine your urethra and bladder for any indications of disease, which can be seen through the lens. Cystoscopy can be done in a doctor's office or a hospital.

- Tissue sampling for scientific research (biopsy). During a cystoscopy, a specialized tool is inserted through the scope of the cystoscope and into your bladder to collect a cell sample for testing. This procedure may also be referred to as "transurethral resection of bladder tumor" (TURBT). TURBT can be used to treat bladder cancer as an alternative treatment option [16].
- Urine samples can be analyzed (urine cytology). Examining a urine sample under a microscope for signs of cancer cells is known as urine cytology. Your doctor can use imaging tests such as a CT urogram or a retrograde pyelogram to examine the structures of your urinary tract. A contrast dye will be injected into a vein in your hand during a CT urogram. Through your ureters and bladder, the dye will eventually pass. In order for your doctor to find any cancerous areas in your urinary tract, the X-ray images taken during the test provide a detailed view of your urinary tract.

7.4 EA FOR TUMOR CLASSIFICATION

Figure 7.1 is a simple model that can be used to diagnose a tumor. It starts with extracting and choosing features, then uses a machine learning model to make predictions. This model takes a tumor image as an input and then works by extracting and choosing features. Boundary detection and genetic operations like disintegration, thresholding, and filtration can be used to pull out features. There are different kinds of image noise in these methods. Because these features were taken out, predictions or diagnoses will be less accurate. If the right features are chosen, the model will be accurate.

Figure 7.1 denotes tumor classification model using evolutionary algorithms. Feature selection is generally performed using genetic algorithms. Classifier or predictor efficiency can be improved and computing complexity reduced by the process of removing factors that have no bearing on the target class. Selecting a subset of features that properly characterize the data in terms of the specified problem space is the primary goal of feature selection. GAs have a long history of research and documentation as effective feature selectors across a wide range of domains. Binary chromosomal bits are a frequent application of GAs as feature selectors, where each bit signifies a

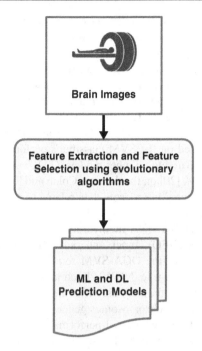

FIGURE 7.1   Tumor classification using EA.

feature's inclusion or exclusion. When calculating fitness, the predictor performance is frequently used as the only metric. Classifiers like Naive Bayes (NB), k-nearest neighbors (k-NN), support vector machines (SVM), decision trees, logistic regression (LR), random forest (RF), or multilayer perceptron networks (MLPN) are typically used in cancer diagnosis [17].

7.4.1 Feature Selection

Ziyad et al. [18, 19, 20] used GAs in wavelet-GA to conduct feature selection on clinical data. Before using the GA, Haar wavelet processing was used to get wavelet coefficients. The chromosomal size was limited to 5 to locate 5 appropriate wavelets. Random sampling using the two-sample t-test filter technique was used to construct the initial population. The fitness function combines average prior distribution and classifier error rate linearly (LDA). The GA used scattered crossover, Gaussian mutations, and stochastic uniform selection. Results from the LDA classifier were compared to those from NB, k-NN, SVM, adaptive resonance theory maps (ARTMAPs), fuzzy ARTMAPs, adaptive network-based fuzzy inference system (ANFIS), and

AdaBoost. Using Wavelet-GA for the ovarian and prostate datasets took a significant amount of time, and the results showed no significant improvement.

It was developed by Sahu et al. [21] using genetic algorithms (GAs) and learning automata (LAs) to select attributes (LA). Patients with various tumors, including leukemia and lymphoma as well as those with a colon tumor, were used in the study. After a mutation, GALA adds a step that is either rewarded or punished by LA. To begin, the genetic algorithm (GA) uses an arbitrary number of genes across all chromosomes to get things going. We calculated fitness based on the SVM classifier's performance on four-fifths of each dataset. Compared to all other models, GALA outperformed SVM + GA [19] and SFS-LDA [20] on the datasets Colon and Tumor 9.

For high-dimensional microarray data related to colon cancer, Jawahar et al. [22] presented a nested-GA based on information from the Cancer Genome Atlas, the TCGA DNA methylation dataset, and the NCBI Gene Expression Omnibus (GEO). Inner and outer GAs make up the nested GA. Every generation of the OGA-SVM requires a lot of computational resources because of IGA-role. NNW's Each layer's output is used to optimise the other layers' initializations. After the first iteration of the IGA-NNW algorithm, the best chromosomes generated by OGA-SVM are used as a starting point. However, nested-performance GA's was erratic when it comes to other datasets. It performed better, equal to or even worse than k-NN and RF, depending on the number of features selected. In the DNA methylation collection, the nested-GA outperformed GA-SVM and GA-NNW for all of the selected genes, but its performance was inconsistent in the other two datasets.

Bhandari et al. [23] presented a two-stage combined approach they called MI-GA for treating colon, ovarian, and lung cancer. Similarity Measure (MI), also known as provisional or uncontrolled entropy, is used to reduce the search space. This option is chosen because of a higher MI score, which indicates less uncertainty. Starting characteristics are randomly distributed across the population by the MI before being fed to the GA. In addition to uniform crossover and mutation and a BCSO, the genetic algorithm (GA) uses a dual and inverse combinator operator [24]. Fitness is determined by SVM accuracy. MI-GA outperformed all other feature selection methods for the five colon kernel functions in an SVM classifier. Although the SVM-polynomial function took longer to run than other functions, it outperformed the others for lung (10 features) and colon (20 features) cancer datasets. Only one strategy seemed to have a clear advantage over ovarian cancer datasets: the linear equation took the most time.

Alhenawi et al. [25] developed a GA-based multilayer feature eliminator (MGRFE) and tested it on 19 microarray cancer datasets. There are three stages to the MGRFE process: random search reduction, wraparound search, and cross-validation. The GA-RFE method is a layer-by-layer selection method. The GA uses techniques like differential integer coding, truncated selection, solitary crossover and mutation to get rid of repetitive genes. The foundation of the training module is the Naive Bayes classifier's accuracy and the adjustments to equilibrium. MGRFE achieved a 100 percent reliability rate in some samples, including some of its earlier models.

SNP Barcodes for breast cancer pathways were discovered using a method devised by Bakos et al. [26]. Hybridization of the Taguchi-genetic algorithm (HTGA). Based on the SNP indices and genotypes found on the chromosomes, a random population was created. After 1000 iterations, the goal was achieved. SNP barcodes with more significant differences between cases and controls were detected by HTGA using the same dataset as particle swarms, chaotic particles and genetic algorithms.

The researchers [27] used GAs to supplement k-NN and SVM-based cancer classification, they used GAs for data reduction and feature selection, respectively. Deconstructing microarray data using DWT allows for the creation of a random population. The concept of fitness is based on the elitism and accuracy of k-NN classifiers. The bit-string mutation is the next step after mono and multipathing crossovers. The features selected were used to train k-NN and SVM classifiers. Using Haar and dp7 DWT, 100 percent accuracy was achieved on four datasets. K-NN achieved 91.67 percent accuracy, and SVM achieved 90 percent accuracy, on the colon dataset. Model selection and classification were done using wavelet-based variable selection and discrete waves for leukemia and colon cancer datasets, respectively.

According to Aziz et al. [28], GAs improved a leukemia classifier model based on microarray data. Using a two-stage feature selection process improves the model's performance based on training genes. An algorithm known as the genetic algorithm (GA) is used on the set of genes chosen to represent a random starting population. The LOOCV values will reveal the degree to which this population is fit that k-NN, SVM, and NB classifiers arrive at for it. The features were analyzed using statistical methods such as k-NN, SVM, NB, linear regression, ANN, and RF. Five of the models achieved accuracy and AUC greater than 0.95, and the relative standard error (RMSE) was low. Results from a comparison of the proposed algorithm to previous research showed that it could achieve 100 percent accuracy with just two genes.

MOGA and XGBoost are used by the researchers [29] to classify cancer. XGBoost, also known as Extreme Gradient Boosting, makes use of Classification and Regression Trees, or CARTS, to assess the importance of various features. MOGA is capable of optimizing a wide range of objective functions under certain conditions. Multi-goal optimization often leads to conflict, making it difficult to solve. When XGBoost was first used, it was for feature selection. The encoded characteristics were then used to build a binary array. A k-NN classifier trained on the chromosomal feature set was used to calculate the fitness score. Crossover and mutation are both accounted for in this model. A 10-fold cross-validation test is then run on 13 cancer datasets to verify the selected features. For the most part, the accuracy of XGBoost0-MOGA features was on par with or slightly better than that of XGBoost, MOGA, or not using any features at all. It was found that both MOGA and XGBoost-MOGA were slower than XGBoost and the other algorithms that were compared. In contrast to XGBoost-MOGA, MOGA performed better.

7.4.2 Parameter Optimization

Bhandari [22] employed GAs for breast cancer feature selection and parameter optimization (GABPNN). Multimodal breast cancer data were localized, standardised, and coded. The fitness value was the reciprocal of the expected error for each chromosome. Each training fold optimises GABPNN network parameters. Against justify feature extraction, the authors evaluate the suggested model to a version of GABPNN without feature selection. However, no benchmarks were used.

Koppad et al. [30] presented a hybrid approach using AdaBoost and GAs to identify tumours from gene expression datasets for UCI repository. The study presents a decision group, a collection of randomly selected classifiers run a set number of times to answer the same issue. The ultimate outcome is voted on by k-NN, Decision Tress, and NB. This decision group is AdaBoost's basis classifier, and the GA optimises their weights. Fitness was a function of each classifier's accuracy. Elitism classifies random single-point crossover and bit-string mutation. AdaBoost-GA was compared against Bagging, RF, Rotation Forest, AdaBoost, AdaBoost-BPNN, AdaBoost-SVM, and AdaBoost-RF. AdaBoost-GA beat all other models in every dataset except lung cancer, when it was second behind AdaBoost-RF. AdaBoost-GA obtained the greatest AUC value for all datasets, although MCC results were inconsistent and other classifiers outperformed. AdaBoost-GA exhibited the lowest variance, suggesting the maximum stability, on colon and brain datasets.

Chen et al. [31] suggested GAs to enhance colorectal cancer image detection. Multiple techniques were used to extract characteristics from RGB photos. Each gene in the GA represented a selection strategy, a classification method, and the amount of characteristics examined for classification. T-statistics, relief algorithm, gain ratio, Information Gain, and chi-squared were used to rank features. k-NN, SVM, MLP, RF, random tree, J48, and KStar were examined. Initial population formation is random within the limits of algorithms and max gene values. The fitness function was the classifier's average AUC value over a certain number of iterations. The models were evaluated and trained on colorectal and NHL datasets, and relief selection was used for all top AUC outcomes. Random Forest had the greatest AUC, while KStar dominated the top 10. For the NHL dataset, k-NN obtained the greatest AUC.

7.5 EA FOR CARCINOMA PREDICTION

Many researchers utilized the genetic algorithms for feature extraction from tumour datasets. Figure shows a model with three parts: feature extraction using GA, feature selection using classifier, and an DL model. Genetic techniques are used to pull out features. Choosing which features to use is the hard part. Learning classifiers have been used a lot in many works of literature.

Evolutionary algorithm-based carcinoma prediction model is shown in Figure 7.2. In many cases, the pre-processed image of the tumor will be used to get the features. In pre-processing, there are two steps: getting rid of image noise and irrelevant pixels.

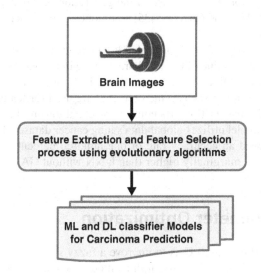

FIGURE 7.2   Carcinoma prediction using EA.

7.5.1 Feature Selection

Using spectroscopy methods, Jian et al. [32] used GAs to uncover key characteristics for classifying bladder cancer. A random selection of six numbers was used to establish the initial population of the SERS. The final classification and the accuracy-based fitness function use linear discriminant analysis (LDA) as a classifier to determine the analysis results. The bottom 25 percent of performers are subjected to single-point crossovers and mutations, while the top 25 percent are selected for additional rounds. The function will be finished after it has completed 100 cycles. A PCA-based LDA classifier was utilised in order to verify the result that was suggested [33]. The PCA model's ROC curve, specificity, and accuracy improved thanks to GAs.

Breast cancer data was utilised by Krishnamurthy et al. [34] to develop cancer detection algorithms that are readable by humans. They did an analysis using the digitised pictures from the Wisconsin Breast Cancer Dataset. In order to discretize the continuous data, a self-organizing map neural network (SOM) is constructed with the help of one hundred samples that are generated at random. The initial population of the GA is composed of random binary vectors. Deterministic reliability is a function of a selection attribute's role in determining fitness. The biological controllers are referred to as single-point crossover and uniform mutation. After 100 iterations of the GA, the characteristics are whittled down with the help of a discoverability vector and put to use in the generation of inducelogical rules. Both the SVM [34] and euro rule [35] techniques were outperformed by the generated rules.

GAs were utilised by Liu et al. [36] in order to select lung cancer-related features. A value of 0 for a characteristic indicates that it is not represented, while a value of 1 indicates that it is. Single-point crossover and bit-string mutation are the two methods that are used for feature selection. The k-NN misclassification procedure is reversed during the optimization process. The model utilizes global data's lung cancer dataset. The proposed method achieved an accuracy rate of one hundred percent for multi-label data, which was marginally higher than k-NN without GA when k was set to 6.

7.5.2 Parameter Optimization

Julio et al. [37] employed GAs to improve a fuzzy system to identify breast cancer. The paper uses the Pittsburgh method, where each genome symbolizes a fuzzy inference system. Randomly produced fuzzy systems are the starting population. The Pittsburgh technique supports combinatorial optimization improvement via variance in the weighting factor. Stochastic uniform selection, single-point crossover, flip-bit mutation.

Davoudi et al. [38] used weighted mean GA to diagnose breast cancer. This study used Breast Cancer Wisconsin data. First, eight classification models are trained, and their accuracy predicted; next, the GA calculates the learning model's weights. SVM [19], decision tree [39], Random Forest [40], Neural Network [41], AdaBoost [42], and Gaussian Nave Bayes [43] are evaluated. The GA is initialized arbitrarily for vertical vector weights. Cross-validation accuracy predicts each classifier's candidate solutions. GA-based weighted predictors were compared to traditional ML models [43]. GA-based models achieved a greater accuracy rate than the classical model.

Houssein et al. [44] used GAs with SVM to enhance breast cancer thermography. The study employed DMR-IR thermal imaging and a university database. The entire method uses two GA ensembles to determine the best modeling and optimized attributes. Uniform crossover, string-bit mutation, and uniform mutation are used to add randomization. The GA is used as a characteristic selector using an arbitrary binary encoded scalar, where 0 denotes not included. The suggested techniques worked well, exceeding in terms of F1-score, accuracy, sensitivity, specificity, and AUC, but the results were not uniform across metrics. Using GA for model selection decreases complexity compared to an exhaustive search, and the model precisely diagnoses breast cancer.

7.6 CONCLUSION

Even though GAs have been utilized effectively in the classification, prediction and diagnosis of several different cancers to a high degree of accuracy, additional study is needed to make them suitable for industrial application. Because they are computationally more expensive than more traditional methods (such as genetic algorithms), GAs are projected to become more affordable as the number of needed evolutions, fitness functions, selection rules, and genetic rules decrease. More recent studies suggest that high accuracy may not be possible for all types of cancer or that different models are needed for each type; Models that can classify multiple types of cancer are urgently needed, as are models for cancer types that have not previously been studied. Classifiers used in combination with GAs may require more investigation.

REFERENCES

1. Mateo, J., Steuten, L., Aftimos, P., André, F., Davies, M., Garralda, E. ... & Voest, E. (2022). Delivering precision oncology to patients with cancer. *Nature Medicine*, *28*(4), 658–665.
2. Cisneros-Villanueva, M., Hidalgo-Perez, L., Rios-Romero, M., Cedro-Tanda, A., Ruiz-Villavicencio, C. A., Page, K. ... & Hidalgo-Miranda, A. (2022). Cell-free DNA analysis in current cancer clinical trials: A review. *British Journal of Cancer*, *126*(3), 391–400.
3. van den Ende, T., van den Boorn, H. G., Hoonhout, N. M., Van Etten-Jamaludin, F. S., Meijer, S. L., Derks, S. ... & van Laarhoven, H. W. (2020). Priming the tumor immune microenvironment with chemo (radio) therapy: A systematic review across tumor types. *Biochimica Et Biophysica Acta (BBA)-Reviews on Cancer*, *1874*(1), 188386.
4. Kim, D. P., Kus, K. J., & Ruiz, E. (2019). Basal cell carcinoma review. *Hematology/Oncology Clinics*, *33*(1), 13–24.
5. Bae, S., Brumbaugh, J., & Bonavida, B. (2018). Exosomes derived from cancerous and non-cancerous cells regulate the anti-tumor response in the tumor microenvironment. *Genes & Cancer*, *9*(3–4), 87.
6. Saleh, T., & Carpenter, V. J. (2021). Potential use of senolytics for pharmacological targeting of precancerous lesions. *Molecular Pharmacology*, *100*(6), 580–587.
7. Singal, A. G., Lampertico, P., & Nahon, P. (2020). Epidemiology and surveillance for hepatocellular carcinoma: New trends. *Journal of Hepatology*, *72*(2), 250–261.
8. Lee, C. K., Man, J., Lord, S., Cooper, W., Links, M., Gebski, V. ... & Yang, J. C. H. (2018). Clinical and molecular characteristics associated with survival among patients treated with checkpoint inhibitors for advanced non–small cell lung carcinoma: A systematic review and meta-analysis. *JAMA Oncology*, *4*(2), 210–216.
9. Kabir, T., Tan, Z. Z., Syn, N. L., Wu, E., Lin, J. D., Zhao, J. J. ... & Goh, B. K. (2022). Laparoscopic versus open resection of hepatocellular carcinoma in patients with cirrhosis: Meta-analysis. *British Journal of Surgery*, *109*(1), 21–29.
10. Paul, S., Rausch, C. R., Jain, N., Kadia, T., Ravandi, F., DiNardo, C. D. ... & Jabbour, E. (2021). Treating leukemia in the time of COVID-19. *Acta Haematologica*, *144*(2), 132–145.
11. Sehn, L. H., & Salles, G. (2021). Diffuse large B-cell lymphoma. *New England Journal of Medicine*, *384*(9), 842–858.
12. Crudele, F., Bianchi, N., Astolfi, A., Grassilli, S., Brugnoli, F., Terrazzan, A. ... & Volinia, S. (2021). The molecular networks of microRNAs and their targets in the drug resistance of colon carcinoma. *Cancers*, *13*(17), 4355.
13. Guida, A., Escudier, B., & Albiges, L. (2018). Treating patients with renal cell carcinoma and bone metastases. *Expert Review of Anticancer Therapy*, *18*(11), 1135–1143.
14. Klein, W. M., Molmenti, E. P., Colombani, P. M., Grover, D. S., Schwarz, K. B., Boitnott, J. ... & Torbenson, M. S. (2005). Primary liver carcinoma arising in people younger than 30 years. *American Journal of Clinical Pathology*, *124*(4), 512–518.
15. Renner, A., Burotto, M., Valdes, J. M., Roman, J. C., & Walton-Diaz, A. (2021). Neoadjuvant immunotherapy for muscle invasive urothelial bladder carcinoma: Will it change current standards? *Therapeutic Advances in Urology*, *13*, 17562872211029779.

16. Iqbal, M. J., Javed, Z., Sadia, H., Qureshi, I. A., Irshad, A., Ahmed, R. ... & Sharifi-Rad, J. (2021). Clinical applications of artificial intelligence and machine learning in cancer diagnosis: Looking into the future. *Cancer Cell International*, *21*(1), 1–11.

17. Ziyad, S. R., Radha, V., & Vaiyapuri, T. (2021). Noise removal in lung LDCT images by novel discrete wavelet-based denoising with adaptive thresholding technique. *International Journal of E-Health and Medical Communications (IJEHMC)*, *12*(5), 1–15.

18. Sahu, B., & Jagadev, A. K. (2021). A GALA based hybrid Gene selection model for identification of relevant genes for cancer microarray data. In Mishra, D., Buyya, R., Mohapatra, P., Patnaik, S. (eds.). *Intelligent and cloud computing* (pp. 827–835). Singapore: Springer.

19. Resmini, R., Silva, L., Araujo, A. S., Medeiros, P., Muchaluat-Saade, D., & Conci, A. (2021). Combining genetic algorithms and SVM for breast cancer diagnosis using infrared thermography. *Sensors*, *21*(14), 4802.

20. Halder, A. K., Delgado, A. H., & Cordeiro, M. N. D. (2022). First multi-target QSAR model for predicting the cytotoxicity of acrylic acid-based dental monomers. *Dental Materials*, *38*(2), 333–346.

21. Jahwar, A., & Ahmed, N. (2021). Swarm intelligence algorithms in gene selection profile based on classification of microarray data: A review. *Journal of Applied Science and Technology Trends*, *2*(01), 01–09.

22. Bhandari, A., Tripathy, B. K., Jawad, K., Bhatia, S., Rahmani, M. K. I., & Mashat, A. (2022). Cancer detection and prediction using genetic algorithms. *Computational Intelligence and Neuroscience*, *2022*.

23. Hidayati, T., Darmawan, E., Indrayanti, I., & Sun, S. (2021). The effect of black cumin seed oil consumption on the platelets and leukocytes number in healthy smokers in rural area Yogyakarta. *Bali Medical Journal*, *10*(3), 1146–1151.

24. Zhang, G., Peng, Z., Yan, C., Wang, J., Luo, J., & Luo, H. (2022). A novel liver cancer diagnosis method based on patient similarity network and DenseGCN. *Scientific Reports*, *12*(1), 1–10.

25. Alhenawi, E. A., Al-Sayyed, R., Hudaib, A., & Mirjalili, S. (2022). Feature selection methods on gene expression microarray data for cancer classification: A systematic review. *Computers in Biology and Medicine*, *140*, 105051.

26. Bakos, B., Kiss, A., Árvai, K., Szili, B., Deák-Kocsis, B., Tobiás, B. ... & Lakatos, P. (2021). Co-occurrence of thyroid and breast cancer is associated with an increased oncogenic SNP burden. *BMC Cancer*, *21*(1), 1–11.

27. YAĞIN, F. H., KÜÇÜKAKÇALI, Z., Cicek, I. B., & BAĞ, H. G. (2021). The effects of variable selection and dimension reduction methods on the classification model in the small round blue cell tumor dataset. *Middle Black Sea Journal of Health Science*, *7*(3), 390–396.

28. Aziz, R. M. (2022). Nature-inspired metaheuristics model for gene selection and classification of biomedical microarray data. *Medical & Biological Engineering & Computing*, *60*(6), 1627–1646.

29. Deng, X., Li, M., Deng, S., & Wang, L. (2022). Hybrid gene selection approach using XGBoost and multi-objective genetic algorithm for cancer classification. *Medical & Biological Engineering & Computing*, *60*(3), 663–681.

30. Koppad, S., Basava, A., Nash, K., Gkoutos, G. V., & Acharjee, A. (2022). Machine learning-based identification of colon cancer candidate diagnostics genes. *Biology*, *11*(3), 365.

31. Chen, X., He, L., Li, Q., Liu, L., Li, S., Zhang, Y. ... & Chen, X. (2022). Non-invasive prediction of microsatellite instability in colorectal cancer by a genetic algorithm–enhanced artificial neural network–based CT radiomics signature. *European Radiology*, 1–12.

32. Jian, Y., Zhang, N., Liu, T., Zhu, Y., Wang, D., Dong, H. ... & Wu, W. (2022). Artificially Intelligent Olfaction for Fast and Noninvasive Diagnosis of Bladder Cancer from Urine. *ACS Sensors*, 7(6), 1720–1731.

33. Taieb, A., Berkovic, G., Haifler, M., Cheshnovsky, O., & Shaked, N. T. (2022). Classification of tissue biopsies by raman spectroscopy guided by quantitative phase imaging and its application to bladder cancer. *Journal of Biophotonics*, e202200009.

34. Krishnamurthy, P., Sarmadi, A., & Khorrami, F. (2021). Explainable classification by learning human-readable sentences in feature subsets. *Information Sciences*, 564, 202–219.

35. Abdullah, D. M., & Abdulazeez, A. M. (2021). Machine learning applications based on SVM classification a review. *Qubahan Academic Journal*, 1(2), 81–90.

36. Liu, B., Yu, H., Zeng, X., Zhang, D., Gong, J., Tian, L. ... & Liu, R. (2021). Lung cancer detection via breath by electronic nose enhanced with a sparse group feature selection approach. *Sensors and Actuators B: Chemical*, 339, 129896.

37. Hernández-Julio, Y. F., Muñoz-Hernández, H., Díaz-Pertuz, L. A., Prieto-Guevara, M., Arrieta-Hernández, N. S., Figueroa-Mendoza, N. A. ... & Nieto-Bernal, W. (2022). Intelligent fuzzy clinical decision support system to predict the Coimbra breast cancer dataset. In Kahraman, C., Tolga, A. C., Cevik Onar, S., Cebi, S., Oztaysi, B., Sari, I. U. (eds.). *International conference on intelligent and fuzzy systems* (pp. 813–821). Cham, Switzerland: Springer.

38. Davoudi, K., & Thulasiraman, P. (2021). Evolving convolutional neural network parameters through the genetic algorithm for the breast cancer classification problem. *Simulation*, 97(8), 511–527.

39. Ghiasi, M. M., & Zendehboudi, S. (2021). Application of decision tree-based ensemble learning in the classification of breast cancer. *Computers in Biology and Medicine*, 128, 104089.

40. Huang, Z., & Chen, D. (2021). A breast cancer diagnosis method based on VIM feature selection and hierarchical clustering random forest algorithm. *IEEE Access*, 10, 3284–3293.

41. Cai, X., Li, X., Razmjooy, N., & Ghadimi, N. (2021). Breast cancer diagnosis by convolutional neural network and advanced thermal exchange optimization algorithm. *Computational and Mathematical Methods in Medicine*, 2021.

42. Yifan, D., Jialin, L., & Boxi, F. (2021, May). Forecast model of breast cancer diagnosis based on RF-AdaBoost. In *2021 International Conference on Communications, Information System and Computer Engineering (CISCE)* (pp. 716–719). Beijing, China: IEEE.

43. Khorshid, S. F., Abdulazeez, A. M., & Sallow, A. B. (2021). A comparative analysis and predicting for breast cancer detection based on data mining models. *Asian Journal of Research in Computer Science*, 8(4), 45–59.

44. Houssein, E. H., Emam, M. M., & Ali, A. A. (2022). An optimized deep learning architecture for breast cancer diagnosis based on improved marine predators algorithm. *Neural Computing and Applications*, 8, 1–19.

Index